BestMasters

Mit „**BestMasters**" zeichnet Springer die besten Masterarbeiten aus, die an renommierten Hochschulen in Deutschland, Österreich und der Schweiz entstanden sind. Die mit Höchstnote ausgezeichneten Arbeiten wurden durch Gutachter zur Veröffentlichung empfohlen und behandeln aktuelle Themen aus unterschiedlichen Fachgebieten der Naturwissenschaften, Psychologie, Technik und Wirtschaftswissenschaften. Die Reihe wendet sich an Praktiker und Wissenschaftler gleichermaßen und soll insbesondere auch Nachwuchswissenschaftlern Orientierung geben.

Springer awards "**BestMasters**" to the best master's theses which have been completed at renowned Universities in Germany, Austria, and Switzerland. The studies received highest marks and were recommended for publication by supervisors. They address current issues from various fields of research in natural sciences, psychology, technology, and economics. The series addresses practitioners as well as scientists and, in particular, offers guidance for early stage researchers.

Melinda Hagedorn

Minimization Problems for the Witness Beam in Relativistic Plasma Cavities

Melinda Hagedorn
Heinrich Heine University Düsseldorf
Düsseldorf, Germany

ISSN 2625-3577   ISSN 2625-3615   (electronic)
BestMasters
ISBN 978-3-658-46225-3   ISBN 978-3-658-46226-0   (eBook)
https://doi.org/10.1007/978-3-658-46226-0

This Springer Spektrum imprint is published by the registered company Springer Fachmedien Wiesbaden GmbH, part of Springer Nature.
The registered company address is: Abraham-Lincoln-Str. 46, 65189 Wiesbaden, Germany

Acknowledgement

During the writing process of this thesis, I received support from a lot of people in many different ways. I would like to take this opportunity to thank these people. First of all, I need to thank Uwe Schulz, whose love and encouragement will always continue influencing me and this world.

Of course, this thesis would not have been possible in its present form without the outstanding supervision I received. During my specialization, Dr. Götz Lehmann patiently supported me. The actual thesis was then predominantly supervised by Dr. Lars Reichwein, whereby Dr. Götz Lehmann and Prof. Florian Jarre were always available to discuss physical and mathematical details. Thank you for providing guidance and constructive feedback throughout this project.

From the bottom of my heart I would like to say big thank you for the love and understanding of my family: Martina Schulz and Axel Schmidt, Michael and Kerstin Hagedorn, Marianne, Hubert and Elke Hagedorn and Jochen Wieschen, Harry, Johanna, Kevin, Lars, Lena, Luna, Marina, Mathias, Sabrina and Sandra Schulz, Andreas, Bernd, Birgit, Michael, Stefan and Dr. Thomas Krakau, Mechthild and Peter Canisius, Claudia and Carolina Eschweiler, Irmgard and Hildegard Herzke, Gabi and Michael Kaldenberg, Ralf Kniffler, Elisabeth and Monika Phillipp, Thorben Schmidt, Harald Schnürch, Alina and Silvia Sporrer. In particular, I would like to thank Dr. Thomas Krakau for carefully proofreading my thesis in addition to my supervisors.

Additionally, I am very grateful for all the inspiring conversations with my brothers and sisters in faith: Laura Ádám Batista, Elisa Berlet, Thomas Bussmann, Dr. Birgit and Dr. Michael Degenfeld, Norbert Fink, Michael Frerichmann, Johannes Gierling, Jonas Grüßem, Marco Gurbisz, Madeleine Hanser,

Stefanie Harmanza, Daniel Heusch, Linda and David Hickson, Theresa Kern, Anne Keßler, Bernadett Kühner, Marion Lechner, Matthew Morrow, Lydia Müller, Torsten Peters, David Scheidler, Christa Schorn, Dennis Schramm, Anna Friederike Six, Lukas Stein, Marie-Therese Strachwitz, Dr. Nils Wiese, Stefan Wißkirchen and Veronika Wollny.

I share memories of learning together and puzzling over exercises with Luke Asbach, Juan Carlos, Florian Dittrich, Marcel Hoffmann, Katharina Krause, Clara Leusch, Oliver Mathiak, Matthias Melchger, Daniel Miller, Thomas Posemann, Giovanni Raso, Diego Santis, Daniel Scherer, Janika Schmidt, Benedikt Schubert and Marius Theißen. Thank you for facing all these challenges together and for being still in contact.

At work, I am and was surrounded by wonderful colleagues who have encouraged me to complete this thesis: Andrea Alexa, Akin Anarat, Dr. Christian Axler, Philipp Berger, Erik Chudzik, Dr. Sina Dahm, Prof. Nils Detering, Bella Duong, Ingo Elsner, Dr. Carsten Feldkamp, Dr. Marina Fischer, Prof. Axel Grünrock, Prof. Christiane Helzel, Dr. Nicole Hufnagel, Dr. Eva-Maria Hüßler, Dr. David Kerkmann, Yanick-Florian Kiechle, Michael Lau, Dr. Felix Lieder, Tobias Löffler, Dr. Christoph Matern, Amelie Porfetye, Dr. Andreas Rätz, Prof. Achim Schädle, Dagmar Schmidt, Prof. Holger Schwender, Dr. Tobias Tietz, Andreas Troll, Joachim Wenk and Kevin Wischnewski.

Although I make an effort to improve and regularly refresh my English skills and have spent one month in Canada, where I was able to learn a lot from my wonderful teachers Laila Lima and David Roth, there is still room for improvement. Nevertheless, I have decided to write this thesis in English. In the following, I will carefully list the tools I used and present transparently how I used them. From time to time, I have had my own English phrases translated into German by www.deepl.com and www.translate.google.com to check that I am being understood correctly. Sometimes I have had this translated back into English to suggest alternative formulations. I used www.de.pons.com and www.collinsdicti onary.com as dictionaries and occasionally googled for synonyms. Furthermore, I used the spell checker that is integrated in "Kile". It is important for me to emphasize that I have not adopted the proposed wordings exactly and am therefore fully responsible for all remaining mistakes.

Contents

List of Figures

List of Tables

Introduction 1

Particle accelerators are used in various applications including basic research in nuclear and particle physics [1], mass spectrometry in chemistry [2] and radiation therapy in medicine [3]. For example, with the Large Hadron Collider (LHC) at CERN [4], protons can be accelerated up to energies of several TeV. A famous example of the groundbreaking successes of particle accelerators is the detection of the Higgs boson at the LHC in 2012 [5]. It had been sought after for decades, since its existence was predicted based on theoretical considerations in 1964 (see [6]). The physicists Peter Higgs and François Englert were awarded the Nobel Prize in 2013.

To increase the energy gain of the accelerated particles the field strength can be increased or the acceleration distance can be extended [7]. Unfortunately, if the electric field strengths are too large, electrical breakdown can occur, which limits this option of improvement. According to [8], the maximum attainable field strength is approximately $100\,\mathrm{MV/m}$. To circumvent this limit, particle accelerators are built very large. The circumference of the LHC, for instance, is almost $27\,\mathrm{km}$. In order to reduce space requirements and costs of future accelerators, approaches are being pursued to accelerate particles using plasma waves.

A promising mechanism is the so-called **laser-driven plasma wakefield acceleration** (LWFA): We assume a gaseous plasma consisting of free electrons and positively charged ions. If a high-intensity laser pulse penetrates the plasma, the electrons are pushed away by the laser pulse predominantly in transversal direction due to the so-called ponderomotive force, which will be derived in Chap. 2. The resulting positive region directly behind the laser pulse re-attracts the displaced electrons. This way, an electronic plasma cavity [9], also known as bubble, is formed. It follows the laser pulse at almost the speed of light [10] and provides strong and uniform accelerating and focusing fields inside.

© The Author(s), under exclusive license to Springer Fachmedien Wiesbaden GmbH, part of Springer Nature 2024

M. Hagedorn, *Minimization Problems for the Witness Beam in Relativistic Plasma Cavities*, BestMasters, https://doi.org/10.1007/978-3-658-46226-0_1

Thus, a high-intensity laser pulse excites electron plasma waves in a plasma, which can be used to accelerate electrons. With high-intensity laser pulses we mean electromagnetic waves with pulse durations of at most 1 ps and intensities of $10^{20}\,\frac{W}{cm^2}$, which corresponds to electric field strengths of approximately 30 TV/m. As described above, plasma can be used to transform this transverse field of the electromagnetic wave efficiently into a longitudinal field that can be used to accelerate particles in the propagation direction of the laser pulse. The nearly harmonic potential that can be used to accelerate charged particles into the direction of propagation provide electric fields of more than 100 GV/m [11].

By using ionization-induced electron injection [12], electrons can be placed inside these strong electric fields inside the bubble in order to be accelerated. Since the injected electrons follow the laser pulse, the electron bunch is also referred to as the witness beam. As it can be seen from Fig. 1.1, the electron bunch is accelerated by the negative region behind the bubble and focused by the electrons surrounding the bubble.

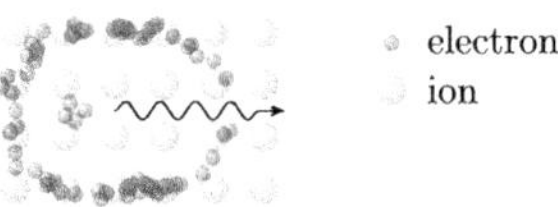

Figure 1.1 Electron bunch in the bubble following the laser pulse

According to [11], the electrons in the witness beam are supposed to repel each other on the one hand, but on the other hand, they are pushed together by the outer electrons. They become arranged in such a way that the described tensions are minimized. The electron configuration is stable in kinetic simulations due to the small probability for Coulomb interactions with plasma background ions [11].

A more precise understanding of this arrangement is very important, because the more regularly the relativistic electrons in the witness beam are structured the higher the spatial coherence. This can be used to realize novel bright sources of short wavelength radiation. For example, the brightness and the spatial coherence of the Compton gamma-source [13] can be improved.

For the mathematical description of the system of the electrons in the witness beam we start in Chap. 3 with the phenomenological bubble model first suggested in [10]. Since the electrons move with a velocity close to the speed of light, the electron-electron interaction of electrons that propagate behind each other is used to be neglected. Nevertheless, at least in case of alongside-propagating electrons with moderate energy the electron interaction needs to be taken into account. This is why [11] added an interaction term to the Hamiltonian of [14]. The Hamiltonian is assumed to describe the total energy of the system of electrons in the witness beam. The electron configuration that minimize the Hamiltonian corresponds to the equilibrium state.

Since the interaction term is calculated from the retarded Liénard-Wiechert potentials, we have to deal with the obstacle that the retarded time is given by an implicit equation. Therefore, [11] approximated the Liénard-Wiechert potentials by a Taylor expansion in terms of v/c, i.e. velocity normalized with the speed of light, and neglected radiation effects. In case of an ultralow-emittance electron bunch, it turns out, that the electrons arrange in hexagonal symmetries, so-called Wigner crystals [15].

A more precise model of the system, which works without Taylor expansions, was developed by [16]. Again, they started with the strongly simplified quasistatic bubble model for electron acceleration in homogeneous plasma [10], which allows to assume the external potentials to be known, and modeled the interaction terms by retarded Liénard-Wiechert potentials. They subdivided the beam load into multiple slices perpendicular to the direction of propagation and assumed that all electrons in the slice are accelerated coherently with much larger kinetic energy than their rest energy. Under these conditions the retarded time can be simplified and computed numerically. This equilibrium slice model for zero transverse emittance beam loads and full Liénard-Wiechert potentials leads to similar hexagonal lattices, but with different size and more topological effects than [11].

In both [11] and [16], the Gradient Descent Method (GDM) is used to minimize the Hamiltonian numerically. One problem are the high computational costs, which increase quadratically with the number of particles. In addition, momenta that are too large result in the minimum possible distances being undercut. In this case, the algorithm does not converge and "NaN" (not a number) is returned.

Therefore, an essential part of this thesis is to examine the suitability of the Gradient Descend Method for our problem and to discuss alternative optimization algorithms. The presented modifications of (GDM) involve free parameters that must be chosen for our problem in such a way that the methods work as well as possible. Finally, the obtained results are compared with each other and with the results of [17]. In this thesis, we restrict ourselves to the two-dimensional case.

Preparation 2

2.1 Basic Information

The aim of this subchapter is to provide information about typical properties of plasma. It is predominantly based on lecture [18] and the books [19], [20], and [21]. Since the different sources use different notations and unit systems, this subchapter is also intended to unify the notation.

2.1.1 What Plasma is and what it can be used for

More than 99% of the observable matter in our universe is made of plasma [20]. This is because luminous stars and the interstellar space, which make up an immense part of the universe, consist mainly of plasma. Furthermore, we can encounter natural plasma on earth itself, for example in the appearance of metal as so-called solid state plasma [22] or as lightning during thunderstorms. Historically, plasma is sometimes referred to as the **fourth state of matter**. It is well known that ice melts into water when heated. If we heat the water further, it evaporates to water vapor. If this water vapor is then heated even more, i.e. the average velocity of the atoms within the water vapor is increased, electrons can be knocked out of the atomic shells of the hydrogen atoms. The gas mixture then contains free electrons and positively charged atoms. This hot, ionized gas is an example of plasma.

In the laboratory, plasma is usually generated by irradiating a gaseous or metallic target with a high-intensity laser pulse [23]. This creates an explosion with an

Supplementary Information The online version contains supplementary material available at https://doi.org/10.1007/978-3-658-46226-0_2.

extremely hot ionized gas in the center for a short period of time: plasma. One problem in dealing with plasma, especially in the context of nuclear fusion, is **confinement** [24]. Our Sun has enough mass to be held together by gravitational confinement [19]. At the same time, the jet pressure generated by the fusion processes inside the Sun ensures that the Sun does not collapse. This dynamically stable state is maintained as long as there is enough fuel [19]. This natural form of confinement cannot be realized in the laboratory. Alternatives are magnetic confinement or inertial confinement, in which the plasma is confined by superconducting magnets or laser pulses, respectively [25, 26].

For example, research is being conducted at the "International Thermonuclear Experimental Reactor" (ITER) to release energy through **nuclear fusion** processes that require the plasma state [27]. As global demand for energy increases and resources such as oil, natural gas, and coal are limited, this is an issue of increasing importance. The current war between Ukraine and Russia again highlighted the lack of alternatives: Wind power and solar energy are dependent on factors that cannot be influenced, which reduces the reliability of these types of energy production. Nuclear power plants, which use nuclear fission, are constantly producing radioactive waste for which a suitable final repository has not been found yet. Nuclear fusion could be the solution, but despite recent progress [28] a lot of research is still required before it can come into everyday use.

A different area of research where plasma approaches are promising is the manipulation of high-intensity laser pulses. If the intensity of a laser pulse is too large, the usually used solid state devices can be damaged. A way out is offered by the following procedure: Two counter-propagating laser beams in underdense plasma can create strong density gratings which exist long enough to change the properties of a third probe pulse. These gratings act as transient plasma photonic crystals (TPPC) and can be used as tunable mirrors [29], frequency filters [30], polarizers and waveplates [31] for laser pulses with intensities beyond the damage threshold of conventional solid state devices.

Another possible field of application for plasma physics is in medicine [32]. Malignant tumors that cannot be approached by surgical methods without damaging vital surrounding areas, for instance some brain tumors, are classically irradiated with high-energy photons. Of course, the treatment can be adjusted to irradiate mainly the tumor. However, because the intensity curve of radiation is continuous, it is not entirely possible to avoid irradiating surrounding healthy tissue as collateral damage as well. The energy spectrum of protons and carbon nuclei, on the other hand, is discrete. They can be accelerated to very high, pretty precise velocities with the help of particle accelerators. Due to their discrete spectrum, protons have a highly precise penetration depth depending on their velocity and the material they penetrate, **proton beams** can be used to destroy inaccessible tumors quite accurately and spare surrounding tissue. This is already being done in Heidelberg and Essen, for example [3]. However, classical particle accelerators are quite large and expensive both to build and to operate. To save costs and space and thus make this treatment available to more people, it would be desirable to accelerate the protons with the help of plasma instead of classical particle accelerators [33]. Accelerating protons using plasma has already been achieved, but the energy and the angular spread of the accelerated protons cannot be controlled precisely enough to make this technique suitable for medical applications. Accordingly, a lot of basic research is still needed to better understand plasma and its properties to become able to use it more effectively in the future.

In order to better distinguish between the different types of plasma (see Sect. 2.1.5), there are some characterizing properties which will be presented here [20]. **Free charge** carriers are characteristic of plasma. These can interact with each other and cause the plasma's good conductivity. In a plasma consisting of positive ions and negatively charged electrons, the degree of ionization is a relevant parameter. Depending on the particle density, the temperature and the ionization energy, a plasma is weaker or stronger ionized. One speaks either of **partially ionized** plasma or fully ionized plasma, in the latter case the atoms have no more bound electrons.

The free charges cause another important property of the plasma: **quasineutrality**. Although a plasma is full of charges interacting with each other via Coulomb forces at the microscopic scale, the free charge carriers can distribute themselves in such a way that the plasma is macroscopically neutral. Free charges are thus able to shield electric fields in the plasma. Viewed from a certain distance, the charge seems to have disappeared. The characteristic length at which the electric potential is smaller than the unshielded Coulomb potential by a factor 1/e is called the **Debye length** λ_D and is defined as:

$$\lambda_D := \sqrt{\frac{\varepsilon_0 k_B T}{e^2 n_0}}, \tag{2.1}$$

where T is the temperature, n_0 the particle density in potential-free space, k_B the Boltzmann constant and ε_0 the permittivity of free space. Note that Euler's number is denoted here by e and the elementary charge by e. The Debye length (2.1) will be derived and discussed in more detail in Sect. 2.1.3.

To talk about plasma the particles have to meet two conditions [21]. The first condition is, that the physical size of the plasma is much larger than the Debye length

$(L \gg \lambda_D)$ and the second condition is, that the product of the particle density and the third power of the Debye length is much larger than unity $(n_0 \cdot \lambda_D^3 \gg 1)$. In this case, the plasma can be called quasi-neutral and collective effects are able to occur.

The next subchapter is dedicated to the so-called plasma frequency. It arises as a basic consequence of Coulomb interactions and is the frequency at which particles in a plasma naturally oscillate. In principle, the corresponding plasma frequency can be determined for all kinds of particles. In this thesis, however, the electron plasma frequency is always meant if the type of particle is not explicitly stated. It is used in Sect. 2.1.7 for normalizations.

2.1.2　Plasma Frequency

An important quantity for the description of a plasma is the plasma frequency ω_p. It indicates the frequency at which an externally applied, time-varying electric field is no longer shielded by the plasma. Since the charge density in the plasma can react to electric fields at most with the plasma frequency, fields that change more slowly than the plasma frequency cannot penetrate the plasma. Thus, electromagnetic waves with a frequency $\omega < \omega_p$ are reflected. See Sect. 2.1.5 for more details.

We consider an electrically neutral plasma in equilibrium [18]. Without an external magnetic field, i.e. $\vec{B} = 0$, one of the Maxwell equations for the electric field $\vec{E}$ and the Lorentz force $\vec{F}_{\mathrm{L}}$ for particles with mass m, charge q, velocity $\vec{v}$, charge density ρ, particle density n and current density $\vec{j} = \rho \vec{v} = qn\vec{v}$ simplify to:

$$\mu_0 \vec{j} + \frac{1}{c^2} \partial_t \vec{E} = \nabla \times \vec{B} = 0, \tag{2.2}$$

$$\vec{F}_{\mathrm{L}} = m\frac{\mathrm{d}\vec{v}}{\mathrm{d}t} = q(\vec{E} + \vec{v} \times \vec{B}) = q\vec{E}, \tag{2.3}$$

where μ_0 denotes the permeability of free space and $\partial_t := \frac{\partial}{\partial t}$. With $c = 1/\sqrt{\varepsilon_0 \mu_0}$ follows

$$\partial_t \vec{E} \stackrel{(2.2)}{=} -c^2 \mu_0 \vec{j} = -\frac{\mu_0}{\varepsilon_0 \mu_0} \rho \vec{v} = -\frac{qn}{\varepsilon_0} \vec{v} \tag{2.4}$$

and using the approximation $\frac{\mathrm{d}}{\mathrm{d}t} = \partial_t + \vec{v} \cdot \nabla \approx \partial_t$ we get

$$\frac{\partial^2 \vec{E}}{\partial t^2} \stackrel{(2.4)}{\approx} -\frac{qn}{\varepsilon_0}\frac{\mathrm{d}\vec{v}}{\mathrm{d}t} \stackrel{(2.3)}{=} -\frac{qn}{\varepsilon_0}\frac{q}{m}\vec{E} = -\frac{q^2 n}{\varepsilon_0 m}\vec{E}. \tag{2.5}$$

This equality can be written as

$$\frac{\partial^2 \vec{E}}{\partial t^2} + \frac{q^2 n}{\varepsilon_0 m}\vec{E} = 0 \tag{2.6}$$

to show that it is the differential equation of a harmonic oscillator with **plasma frequency** ω_p and therefore especially for electrons $\omega_{p,e}$ and for ions $\omega_{p,i}$:

$$\omega_p = \sqrt{\frac{q^2 n}{\varepsilon_0 m}}, \quad \omega_{p,e} = \sqrt{\frac{e^2 n_e}{\varepsilon_0 m_e}}, \quad \omega_{p,i} = \sqrt{\frac{e^2 n_i}{\varepsilon_0 m_i}}, \tag{2.7}$$

where n_i and m_i denote the particle density and the mass of the ions.

2.1.3   Debye Shielding

As announced in Sect. 2.1.1, we will take a closer look at quasi-neutrality caused by Debye shielding. For this purpose, we assume a plasma consisting of electrons and ions with electric charge $+e$. We assume that the ions are significantly heavier than the electrons, so that they can be regarded as static on the time scale under consideration.

If we now introduce a sample charge q_0 into our plasma, it can be shielded by the unbound electrons. The trajectories of the very fast moving electrons are deflected by the sample charge, so that Figure 2.1 can be interpreted as a snapshot. Our aim is to derive the potential ϕ of this sample charge following the derivation in [21].

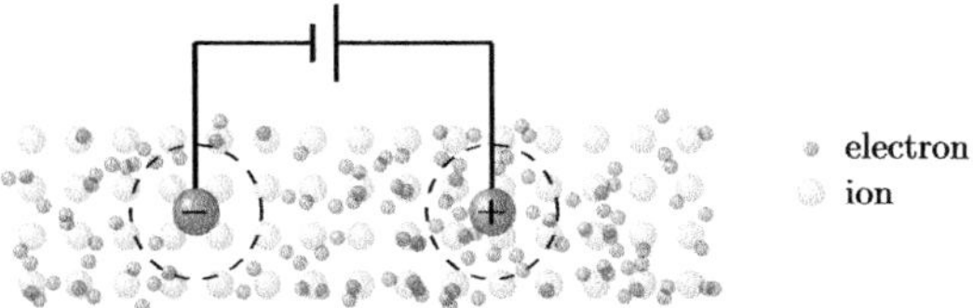

Figure 2.1 The free electrons in a plasma are attracted by positive sample charges and repelled by negative sample charges. Outside the Debye radius, the potentials are therefore shielded, i.e. barely perceptible. Ions are assumed to be static

The charge density ρ is in this case given by

$$\rho = e(n_i - n_e) + q_0\frac{\delta(r)}{r^2}, \tag{2.8}$$

where the ion density n_i remains approximately the unperturbed plasma density n_0 due to the high mass of the ions. The δ distribution models the point-like density of the sample charge. Assuming a thermal equilibrium, the electrons can be described by the Boltzmann distribution and thus, the electron density n_e can be written as

$$n_e = n_0 \exp\left(-\frac{q\phi}{k_B T}\right) = n_0 \exp\left(\frac{e\phi}{k_B T}\right), \tag{2.9}$$

with electric potential ϕ and electron temperature T. If $|e\phi| \ll k_B T$, a Taylor expansion of the Boltzmann factor is justified:

$$\exp\left(\frac{e\phi}{k_B T}\right) = \sum_{k=0}^{\infty} \frac{1}{k!}\left(\frac{e\phi}{k_B T}\right)^k = 1 + \frac{e\phi}{k_B T} + \mathcal{O}\left(\left(\frac{e\phi}{k_B T}\right)^2\right) \approx 1 + \frac{e\phi}{k_B T}. \tag{2.10}$$

Let $r \neq 0$ denote the distance to the sample charge. Due to the definition of the Laplace operator in spherical coordinates, the Poisson equation leads to

$$\frac{1}{r^2}\partial_r(r^2\partial_r)\phi = \Delta\phi = -\frac{\rho}{\varepsilon_0} \overset{(2.8)}{=} -\frac{e}{\varepsilon_0}(n_i - n_e) \overset{(2.9),(2.10)}{\approx} -\frac{e}{\varepsilon_0}\left(n_0 - n_0\left(1 + \frac{e\phi}{k_B T}\right)\right)$$

$$= \frac{e}{\varepsilon_0}n_0\frac{e\phi}{k_B T} = \frac{n_0 e^2}{\varepsilon_0 k_B T}\phi \overset{(2.1)}{=} \frac{1}{\lambda_D^2}\phi. \tag{2.11}$$

In order to find a solution of this differential equation we perform several steps. First, we simplify the left hand side by using the ansatz $\phi(r) = \frac{f(r)}{r}$:

$$\frac{1}{r^2}\partial_r(r^2\partial_r)\phi(r) = \frac{1}{r^2}\partial_r\left(r^2\partial_r\frac{f(r)}{r}\right) = \frac{1}{r^2}\partial_r\left(r^2\left(\frac{-f(r)}{r^2} + \frac{\partial_r f(r)}{r}\right)\right)$$

$$= \frac{1}{r^2}\partial_r(-f(r) + r\cdot\partial_r f(r)) = \frac{1}{r^2}\left(-\partial_r f(r) + \partial_r f(r) + r\partial_r^2 f(r)\right)$$

$$= \frac{1}{r}\partial_r^2 f(r) \overset{!}{=} \frac{1}{\lambda_D^2}\frac{1}{r}f(r) = \frac{1}{\lambda_D^2}\phi(r).$$

The second step is solving the differential equation $\partial_r f(r) = \frac{1}{\lambda_D^2}f(r)$ that we have received as a result. It is solved by functions of the form $f(r) = Ae^{r/\lambda_D} + Be^{-r/\lambda_D}$ for arbitrary constants A and B.

Finally, we consider the two boundary conditions. On the one hand, we require that the potential vanishes infinitely far away from the sample charge, i. e.

$$\lim_{r \to \infty} \phi(r) = \lim_{r \to \infty} \frac{1}{r}\left(A e^{r/\lambda_D} + B e^{-r/\lambda_D}\right) = \lim_{r \to \infty} \frac{1}{r} A e^{r/\lambda_D} \overset{!}{=} 0, \qquad (2.12)$$

which gives us $A \equiv 0$. On the other hand, the potential we are looking for should match the Coulomb potential close to the sample charge. Thus, for small r we need

$$\phi(r) = \frac{1}{r} B e^{-r/\lambda_D} \approx \frac{1}{r} B \overset{!}{=} \frac{q_0}{4\pi \varepsilon_0 r} = \phi_{\text{Coulomb}}(r), \qquad (2.13)$$

leading to $B \equiv q_0/(4\pi \varepsilon_0)$. All in all, we derived the Debye-Hückel-Potential

$$\phi(r) = \frac{q_0}{4\pi \varepsilon_0 r} e^{-r/\lambda_D}. \qquad (2.14)$$

As it can be seen from Figure 2.2, the Debye-Hückel-Potential asymptotically approaches the Coulomb potential for small distances and zero for large distances. In particular, it drops significantly faster than the Coulomb potential and vanishes shortly after the Debye length λ_D. These observations are in accordance with our analytic formula and our boundary conditions.

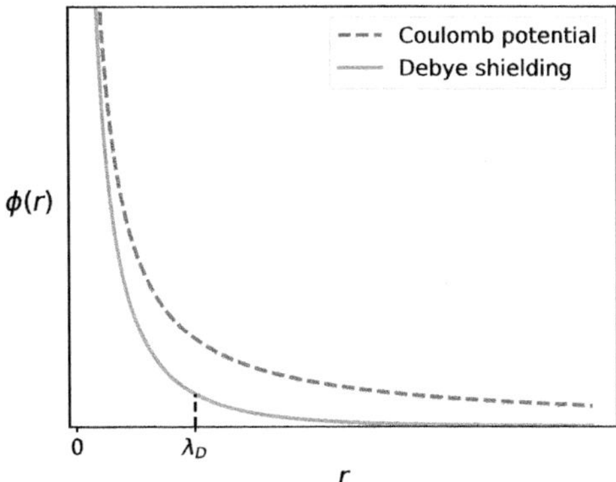

Figure 2.2 Coulomb potential and Debye-Hückel-Potential

The Debye-Hückel-Potential (2.14) allows us to interpret the Debye length λ_D as the distance, at which the potential is only $\frac{1}{e}$ times the Coulomb potential. It defines the order of magnitude in which the influence of a sample charge should definitely be taken into account. If the distance from the sample charge is significantly larger than λ_D, the effects are negligible. Note, that this characteristic distance is independent of the specific sample charge.

2.1.4 Wave Equation

Before we analyze the more complicated interactions between a laser pulse and plasma, we first take a look at electromagnetic waves in a vacuum. With the help of the Maxwell equations we derive the wave equation. The simplest solution of the wave equation are plane waves. They lead to the dispersion relation in vacuum, which will be compared to the dispersion relation in plasma in Sect. 2.1.5.

Unfortunately, plane waves have infinite extension. Thus, the ponderomotive force, which will be derived and discussed in Sect. 2.2, could not displace electrons, when a laser pulse penetrates a plasma. But it does, as addressed in Chap. 1 and will be discussed in more detail in Chapter 3. A spatially restriction is therefore absolutely necessary to correctly describe the processes taking place. We use another solution of the wave equation, spherical waves, to derive the Gaussian beam, which is only an approximate solution of the wave equation, but more realistic.

In vacuum there is no charge and no current, i.e. $\rho = 0$ and $\vec{j} = 0$. Thus, the Maxwell's equations simplify to

(i) $\nabla \cdot \vec{E} = 0,$
(ii) $\nabla \cdot \vec{B} = 0,$
(iii) $\nabla \times \vec{E} = -\frac{\partial \vec{B}}{\partial t},$
(iv) $\nabla \times \vec{B} = \frac{1}{c^2} \frac{\partial \vec{E}}{\partial t}.$

In addition, we need the following relation:

Proposition 2.1 *For any* $\vec{E} := \vec{E}(\vec{r})$ *the following is true:*

$$\nabla \times (\nabla \times \vec{E}) = \nabla(\nabla \cdot \vec{E}) - \nabla \vec{E}.$$

Proof. Consider the i-th component of $\nabla \times (\nabla \times \vec{E})$, where $i \in \{1, 2, 3\}$. The Levi-Civita symbol is denoted by ε_{ijk}, the Laplace operator by $\Delta := \nabla^2 = \partial_x^2 + \partial_y^2 + \partial_z^2$ and the Kronecker delta by $\delta_{i,j}$. Note the Einstein summation convention and Schwarz's theorem (symmetry of second derivatives), then

$$(\nabla \times (\nabla \times \vec{E}))_i = \varepsilon_{ijk}(\partial_j(\nabla \times \vec{E})_k) = \varepsilon_{ijk}\,\partial_j\,\varepsilon_{k\ell m}\,\partial_\ell\,E_m = \varepsilon_{ijk}\,\varepsilon_{k\ell m}\,\partial_j\,\partial_\ell\,E_m$$

$$= (\delta_{i,\ell}\,\delta_{j,m} - \delta_{i,m}\,\delta_{j,\ell})\partial_j\,\partial_\ell\,E_m = \partial_i\partial_j E_j - \partial_j\partial_j E_i = (\nabla(\nabla \cdot \vec{E}) - \Delta\vec{E})_i$$

is true for every component and thus the assertion follows. $\qquad\square$

Together with

$$\nabla \times (\nabla \times \vec{E}) \overset{(iii)}{=} \nabla \times \left(-\frac{\partial \vec{B}}{\partial t}\right) = -\frac{\partial}{\partial t}(\nabla \times \vec{B}) \overset{(iv)}{=} -\frac{\partial}{\partial t}\frac{1}{c^2}\frac{\partial \vec{E}}{\partial t} = \frac{-1}{c^2}\frac{\partial^2 \vec{E}}{\partial t^2}$$

$$(2.15)$$

this proposition can be used to derive the **wave equation**

$$\frac{-1}{c^2}\frac{\partial^2 \vec{E}}{\partial t^2} \overset{(2.15)}{=} \nabla \times (\nabla \times \vec{E}) = \nabla(\nabla \cdot \vec{E}) - \Delta\vec{E} \overset{(i)}{=} -\Delta\vec{E}, \qquad (2.16)$$

which can also be written in the form

$$\Box\vec{E} = \left(\Delta - \tfrac{1}{c^2}\partial_t^2\right)\vec{E} = 0, \qquad (2.17)$$

where $\Box := \Delta - \frac{1}{c^2}\partial_t^2$ is called **d'Alembert operator**, which is also known as wave operator or quabla operator (see [34]).

A dispersion relation generally describes the relationship between the course and the properties of a physical process. In the context of electromagnetic waves the process can be described, for instance, with the help of the frequency ω or the energy and the properties with the wavenumber k or the wavelength λ. Electromagnetic waves are solutions of the wave equation.

Proposition 2.2 *One solution of the wave equation is the **plane wave** that can be written as*

$$\vec{E}(\vec{r}, t) = \vec{E}_0 \cos(\omega t - \vec{k}^T\vec{r}) = \tfrac{1}{2}\vec{E}_0 e^{i(\vec{k}^T\vec{r}-\omega t)} + \text{c.c.}, \qquad (2.18)$$

where $\vec{k}$ is a vector in the direction of propagation, $\vec{E}_0$ the constant amplitude and "c.c." stands for "complex conjugate".

Proof. Substituting the plane wave into the wave equation yields

$$\Box \vec{E}(\vec{r}, t) = \left(\Delta - \tfrac{1}{c^2}\partial_t^2\right) \vec{E}_0 \cos(\omega t - \vec{k}^{\mathsf{T}}\vec{r}) = (-k^2 - \tfrac{1}{c^2}(-\omega^2))\vec{E}(\vec{r}, t)$$
$$= (\tfrac{\omega^2}{c^2} - k^2)\vec{E}(\vec{r}, t) = 0,$$

which leads to the **dispersion relation** $k^2 = \tfrac{\omega^2}{c^2}$. $\qquad\qquad\qquad\qquad\qquad\square$

The following discussion is oriented to [35]. The electric field $\vec{E}(\vec{r}, t)$ is maximized, when it holds $\cos(\omega t - \vec{k}^{\mathsf{T}}\vec{r}) = 1$, i. e. when $\omega t - \vec{k}^{\mathsf{T}}\vec{r} = -2\pi m$ for $m \in \mathbb{Z}$.

In the one-dimensional case $\vec{r} \equiv (0, 0, z)^{\mathsf{T}}$ this condition simplifies to $\omega t - kz = -2\pi m$, which means that the maxima are at

$$z = (\omega t + 2\pi m)/k = \tfrac{\omega}{k}t + \tfrac{2\pi}{k}m = \pm ct + m\lambda. \qquad (2.19)$$

This implies that the maxima propagate with the speed of light and have the distance λ to each other. The problem with plane waves is that an infinite extension is assumed. Another solution of the wave equation leads to spherical waves

$$\vec{E}(\rho) = \tfrac{\vec{A}}{\rho}e^{-i(k\rho - \omega t)} + \text{c.c.} \qquad (2.20)$$

with amplitude $\vec{A}$ and the radius from the origin of the wave $\rho = \sqrt{x^2 + y^2 + z^2}$. It is assumed that they propagate in every direction. Both types of waves are therefore not suitable for adequately modeling laser beams.

A more accurate model is a plane wave bounded by an aperture. According to the Huygens-Fresnel principle, each point of the aperture radiates a spherical wave. A smaller divergence angle is obtained if the intensity profile is initially given by a local Gaussian distribution. Since the wave equation is invariant under the transformation $z \to z + iz_R =: q$, spherical waves with imaginary center

$$\vec{E}(r, z, t) = \frac{\vec{A}}{\sqrt{q^2 + r^2}}e^{-i\left(k\sqrt{q^2 + r^2} - \omega t\right)} \qquad (2.21)$$

also solve the wave equation, where $r := \sqrt{x^2 + y^2}$.

Splitting $1/q$ into a real and an imaginary part leads to

$$\frac{1}{q} = \frac{1}{z + iz_R} = \frac{z - iz_R}{z^2 + z_R^2} = \frac{z}{z^2 + z_R^2} - i\frac{z_R}{z^2 + z_R^2} = \frac{1}{R(z)} - i\frac{2}{k\omega(z)^2} \qquad (2.22)$$

with ray radius

$$\omega(z) := \sqrt{2(z^2 + z_R^2)/(kz_R)} \qquad (2.23)$$

and radius of curvature

$$R(z) = (z^2 + z_R^2)/z. \qquad (2.24)$$

Close the z-axis, $r \ll q$ is valid and (2.21) can be approximated as

$$
\begin{aligned}
\vec{E}(r, z, t) &\overset{(2.21)}{\approx} \frac{\vec{A}}{q} \exp\left(-i\left(kq\sqrt{1 + \left(\tfrac{r}{q}\right)^2} - \omega t\right)\right) \approx \frac{\vec{A}}{q} \exp\left(-i\left(kq\left(1 + \tfrac{r^2}{2q^2}\right) - \omega t\right)\right) \\
&= \frac{\vec{A}}{q} \exp\left(-i\left(kz + ikz_R + \tfrac{kr^2}{2q} - \omega t\right)\right) = \frac{1}{q} \underbrace{\vec{A}e^{kz_R}}_{=: \tilde{A}} \exp\left(-i\tfrac{kr^2}{2q}\right) \exp\left(i(\omega t - kz)\right) \\
&= \frac{\tilde{A}}{q} \exp\left(-i\tfrac{kr^2}{2q}\right) \exp\left(i(\omega t - kz)\right) = \frac{\tilde{A}}{q} \exp\left(-i\tfrac{kr^2}{2R(z)}\right) \exp\left(-\tfrac{r^2}{\omega(z)^2}\right) \exp\left(i(\omega t - kz)\right)
\end{aligned}
$$

with the help of the Taylor expansion

$$\sqrt{1 + x} \approx 1 + \tfrac{1}{2}x \qquad (2.25)$$

for small x and with amplitude $\tilde{\vec{A}} := \vec{A}e^{kz_R}$. This approximate solution of the wave equation is called **Gaussian beam**.

2.1.5 Classification

It is necessary to distinguish between non-relativistic and **relativistic** plasma [20]. The velocity-threshold when to use relativistic calculations depends on the specific problem. For example, in the context of gases the rough estimation $v_{th}^2 \geq 0.05c^2$ can be used, i.e. relativistic effects have at least to be taken into account if the square of the most probable, thermal velocity [36] of the electrons $v_{th} = \sqrt{2k_B T_e/m_e}$ is larger than 5% of the speed of light in vacuum squared. Here, m_e denotes the electron mass and k_B the Boltzmann constant. This corresponds to an electron temperature of

$$T_e = v_{th}^2 \frac{m_e}{2k_B} \gtrsim 0.05c^2 \cdot \frac{m_e}{2k_B} \approx 1.5 \cdot 10^8 \, \text{K}. \tag{2.26}$$

Furthermore, a differentiation must be made between **quantum** plasma and classical plasma [20]. If the particle density is large in relation to the temperature, then the Fermi principle has to be considered. As a rough estimate, one can speak of classical plasma, if the de Broglie wavelength $\lambda_{dB} = \frac{h}{p}$ with momentum p and Planck constant h is significantly smaller than the mean distance of the electrons, which can be approximated by $n_e^{-1/3}$. Thus, quantum effects must be taken into account if

$$n_e \gtrsim (\lambda_{dB})^{-3} = \left(\frac{h}{p}\right)^{-3} = \left(\frac{h}{m_e v_{th}}\right)^{-3} = \left(\frac{h}{m_e}\sqrt{\frac{m_e}{2k_B T_e}}\right)^{-3} = \left(\frac{h^2}{2m_e k_B}\right)^{-3/2} T_e^{3/2}. \tag{2.27}$$

A plasma is called **ideal** if Coulomb collisions can be neglected [20]. This is the case when the potential energy E_{pot}, which is proportional to $n_e^{1/3}$, is very small compared to the kinetic energy E_{kin}. In other words, for an ideal plasma the following must be true:

$$\frac{E_{\text{pot}}}{E_{\text{kin}}} \ll 1. \tag{2.28}$$

Now, for which densities this is the case depends on whether we are dealing with a classical or a quantum plasma. In a classical, gaseous plasma, the kinetic energy of the electrons is proportional to $k_B T_e$. Thus

$$\frac{E_{\text{pot}}}{E_{\text{kin}}} \propto n_e^{1/3} T_e^{-1} \tag{2.29}$$

and one can speak of an ideal plasma in the case of classical plasma if the density is particularly low. If, on the other hand, a quantum plasma is present, it applies $E_{\text{kin}} = \frac{p^2}{2m_e} = \frac{h^2}{2m_e \lambda_{dB}^2}$ and therefore

$$\frac{E_{\text{pot}}}{E_{\text{kin}}} \propto \frac{n_e^{1/3}}{n_e^{2/3}} = n_e^{-1/3}, \tag{2.30}$$

which means that a quantum plasma may be assumed to be an ideal plasma if the density is particularly high.

Next, we take a look at the **dispersion relation**. In plasma, this is given by

$$\omega^2 = \omega_p^2 + k^2 c^2. \tag{2.31}$$

A derivation of this formula can be found in [21]. Although the approximation of the cold plasma is used there, this does not represent a major limitation for the applicability of the theory, since essential aspects of wave propagation can be transferred to hot plasma. For low densities n_e, i. e. for plasma frequencies $\omega_p \propto \sqrt{n_e} \to 0$, and for high frequencies $\omega \to \infty$, these waves become asymptotically vacuum waves known from Proposition 2.2 in Sect. 2.1.4. A sketch of the plasma dispersion relation (2.31) is given in Figure 2.3.

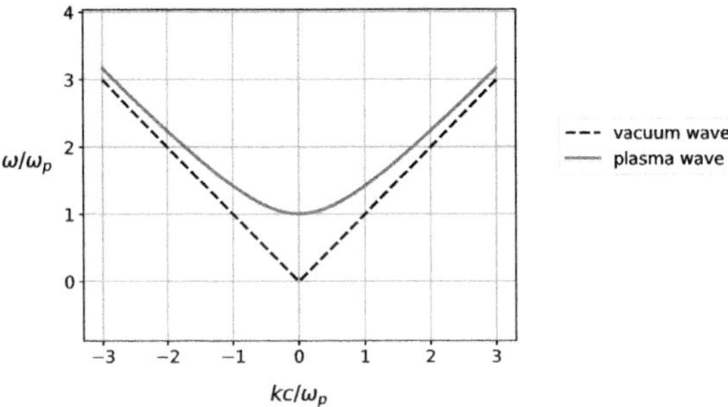

Figure 2.3 Dispersion relation in plasma and in vacuum

From this dispersion relation we can deduce two important quantities: The phase velocity

$$v_{ph} := \frac{\omega}{k}, \tag{2.32}$$

which is the velocity of a single-frequency component, and the group velocity

$$v_g := \frac{\partial \omega}{\partial k}, \tag{2.33}$$

defined as the velocity of the center of a pulse (see [37]). Since the frequencies ω and ω_p are real numbers, the **phase velocity** in plasma is larger than the speed of light:

$$v_{ph} = \frac{\omega}{k} \overset{(2.31)}{=} \frac{\sqrt{\omega_p^2 + k^2 c^2}}{k} = \sqrt{\frac{c^2 \omega_p^2}{c^2 k^2} + c^2} = c\sqrt{\frac{\omega_p^2}{c^2 k^2} + 1}$$

$$= c\sqrt{\frac{\omega_p^2}{\omega^2 - \omega_p^2} + \frac{\omega^2 - \omega_p^2}{\omega^2 - \omega_p^2}} = c\sqrt{\frac{\omega^2}{\omega^2 - \omega_p^2}} > c. \tag{2.34}$$

This is not a contradiction to Einstein's theory of relativity, since no energy or information is transferred at phase velocity. This happens with the group velocity, which, as we will see later in this chapter, is less than the speed of light. Using the derived phase velocity, we are able to find a formula for the **refractive index**

$$N = \frac{c}{v_{ph}} \overset{(2.34)}{=} \sqrt{\frac{\omega^2 - \omega_p^2}{\omega^2}} = \sqrt{1 - \frac{\omega_p^2}{\omega^2}}. \tag{2.35}$$

From this formula we see immediately that in case of $\omega > \omega_p$ we get a refractive index in the interval $]0, 1[$. The limit case $\omega = \omega_p$ leads to $N = 0$ and we observe reflection. The case $\omega < \omega_p$ would lead to a purely imaginary refractive index and therefore laser pulses with lower frequencies than the plasma frequency are not able to propagate through the considered plasma. Thus, when $\omega > \omega_p$, energy can be transferred with **group velocity**

$$v_g = \frac{\partial \omega}{\partial k} \overset{(2.31)}{=} \frac{\partial}{\partial k}\sqrt{k^2 c^2 + \omega_p^2} = \frac{1}{2}\frac{2kc^2}{\sqrt{k^2 c^2 + \omega_p^2}} = \frac{c}{\sqrt{1 + \frac{\omega_p^2}{k^2 c^2}}} \tag{2.36}$$

$$= \frac{c}{\sqrt{\frac{\omega^2 - \omega_p^2}{\omega^2 - \omega_p^2} + \frac{\omega_p^2}{\omega^2 - \omega_p^2}}} = c\sqrt{\frac{\omega^2 - \omega_p^2}{\omega^2}} = c\sqrt{1 - \frac{\omega_p^2}{\omega^2}} < c, \tag{2.37}$$

in accordance with Einstein's theory of relativity.

From a slightly different point of view, we are able to distinguish between underdense and overdense plasma for a given laser pulse [38]. Assuming $\omega = \omega_p$, we get from Equation (2.7) the **critical density**

$$n_c = \frac{\varepsilon_0 m_e}{e^2}\omega^2. \tag{2.38}$$

Imagine a plasma with increasing electron density n_e in the direction of propagation of the laser pulse (see Figure 2.4). As long as $n_e < n_c$, the laser pulse can propagate through the plasma and the plasma is called **underdense** [39]. If the laser pulse reaches the critical density, it is reflected. Strictly speaking, the wave has still a finite amplitude up to the penetration depth δ behind the cutoff density n_c, analogously to the tunnel effect. Beyond that, the (non-relativistic) wave can no longer propagate in the plasma, since $n_e > n_c$ leads to a purely imaginary refractive index as discussed before. In this case, we call the plasma **overdense**.

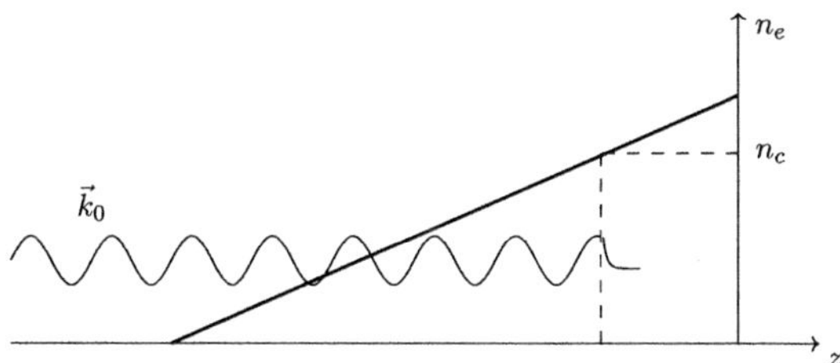

Figure 2.4 Plasma with increasing electron density n_e in z-direction. Reflection at cutoff density n_c after penetration depth δ

As already hinted, under certain conditions relativistically intense laser beams can propagate in overdense plasma [40]. One mechanism is that the self-focusing laser beam digs a hole in the overdense plasma. The other mechanism is the high electron quivering velocity.

2.1.6   Distribution Function

The electron-ion plasma mentioned in Chap. 1 consists of a huge number of small, similar particles. Instead of considering the position of each individual particle separately, it is common to consider a particle type as one batch whose particles have their individual properties with a certain probability. A batch of particles is described completely by its distribution function.

For example, the Boltzmann distribution was already used in Sect. 2.1.3 to describe the electron density. We therefore do not know the exact position of a particular particle, but we can predict how likely it is to encounter a particle at a particular location.

Laser pulses can also be described using distribution functions, for example using the normal distribution. Chapter 2.1.8 explains how the underlying distribution of a laser pulse is related to its so-called pulse duration.

The distribution function $f := f(\vec{r}, \vec{v}, t)$ is nothing else than the probability density in the six-dimensional phase space [18]. It characterizes every species of particles by its position $\vec{r}$ and velocity $\vec{v}$ with $v := |\vec{v}|$ at the time t and can be used to determine

- the total number of particles

$$N = \int_{\mathbb{R}^3} \mathrm{d}^3\vec{r} \int_{\mathbb{R}^3} \mathrm{d}^3\vec{v}\ f(\vec{r}, \vec{v}, t), \qquad (2.39)$$

- the local density

$$n(\vec{r}) = \int_{\mathbb{R}^3} \mathrm{d}^3\vec{v}\ f(\vec{r}, \vec{v}, t), \qquad (2.40)$$

- the current density

$$\vec{j}(\vec{r}) = \int_{\mathbb{R}^3} \mathrm{d}^3\vec{v}\ q\vec{v} f(\vec{r}, \vec{v}, t), \qquad (2.41)$$

- and the kinetic energy density

$$\varepsilon(\vec{r}) = \int_{\mathbb{R}^3} \mathrm{d}^3\vec{v}\ \frac{mv^2}{2} f(\vec{r}, \vec{v}, t). \qquad (2.42)$$

To describe a collisionless plasma consisting of charged particles, we can utilize the total derivative of the distribution function with respect to time, which is required for the **Vlasov equation**

$$\frac{\mathrm{d}f}{\mathrm{d}t} = 0. \qquad (2.43)$$

This kinetic equation provides the most general description of plasma [18]. The zero on the right-hand side means that the particle flux in the phase space is incompressible. Using the Lorentz force $\vec{F}_\mathrm{L} = \frac{\mathrm{d}\vec{p}}{\mathrm{d}t} = \dot{\vec{p}} = m\dot{\vec{v}} = q(\vec{E} + \vec{v} \times \vec{B})$ with momentum $\vec{p}$, mass m, charge q, electric field $\vec{E}$ and magnetic field $\vec{B}$, the Vlasov equation can be written as

$$\frac{\mathrm{d}f}{\mathrm{d}t} = \frac{\partial f}{\partial t} + \frac{\partial f}{\partial \vec{r}}\frac{\mathrm{d}\vec{r}}{\mathrm{d}t} + \frac{\partial f}{\partial \vec{v}}\frac{\mathrm{d}\vec{v}}{\mathrm{d}t} = \frac{\partial f}{\partial t} + \vec{v}\nabla f + \frac{q}{m}(\vec{E} + \vec{v} \times \vec{B})\frac{\partial f}{\partial \vec{v}} = 0.$$
$$(2.44)$$

2.1.7 Natural Units

Especially in programming it is reasonable to eliminate units. For this purpose, a variable is divided into a dimensionless number and a unit part. The unit part depends on physical constants like the vacuum speed of light c, the elementary charge e or the electron mass m_e. For example, the electric charge q can be split in the following manner

$$q = \hat{q} \cdot q_{\text{unit}} \tag{2.45}$$

with dimensionless number $\hat{q}$ and unit part $q_{\text{unit}} = e$. The advantage of this procedure is that the charge of an electron can be written dimensionless as

$$\hat{q}_e = \frac{q_e}{q_{\text{unit}}} = \frac{-e}{e} = -1. \tag{2.46}$$

In a similar way, other dimensionless quantities can be derived. With speed of light c, electron mass m_e, elementary charge e and plasma frequency ω_p we get for the following quantities (Table 2.1):

Table 2.1 Overview of natural units

Quantity	Natural unit
Charge q	$q_{\text{unit}} = e$
Time t	$t_{\text{unit}} = \frac{1}{\omega_p}$
Length x	$x_{\text{unit}} = \frac{c}{\omega_p}$
Velocity v	$v_{\text{unit}} = c$
Mass m	$m_{\text{unit}} = m_e$
Momentum p	$p_{\text{unit}} = m_e c$
Scalar potential ϕ	$\phi_{\text{unit}} = \frac{m_e c^2}{e}$
Vector potential A	$A_{\text{unit}} = \frac{m_e c}{e}$
Electric field E	$E_{\text{unit}} = \frac{m_e \omega_p c}{e}$
Magnetic field B	$B_{\text{unit}} = \frac{m_e \omega_p}{e}$
Energy $\mathcal{E}$	$\mathcal{E}_{\text{unit}} = m_e c^2$

2.1.8 Full Width at Half Maximum

As mentioned in Chap. 1, in LWFA we are dealing with laser pulses with a very short pulse duration. However, since laser pulses are described using distribution functions that converge to zero at infinity but are never exactly zero from a mathematical point of view, this way of speaking can be somewhat confusing. In this subsection, the normal distribution and a Gaussian pulse are used to illustrate what is meant by a pulse duration or pulse width. In other areas such as spectroscopy [41] or in the context of solitons in optics [42], other distribution functions such as the Cauchy distribution or the hyperbolic secant are more usual. However, the procedure presented here is transferable.

In a distribution $f(x)$, we consider the maximum function value $f_{\max}$. For the above mentioned functions, the half of this value, i. e. $\frac{f_{\max}}{2}$, is taken exactly twice, at the positions $x_{1/2}^{(1)}$ and $x_{1/2}^{(2)}$. The absolute value of the difference of these two values $\left| x_{1/2}^{(1)} - x_{1/2}^{(2)} \right|$ is called **full width at half maximum (FWHM)**.

Consider, for example, the normal distribution with density

$$f(x) = \frac{1}{\sqrt{2\pi\sigma^2}} e^{-\frac{1}{2\sigma^2}(x-x_0)^2}, \tag{2.47}$$

where $x_0 \in \mathbb{R}$ is the expected value and $\sigma^2 > 0$ the variance. The maximum is at x_0 and has the value $f(x_0) = \frac{1}{\sqrt{2\pi\sigma^2}}$. In order to find $x_{1/2}^{(1)}$ and $x_{1/2}^{(2)}$, we calculate for $j \in \{1, 2\}$:

$$f(x_{1/2}^{(j)}) = \frac{1}{\sqrt{2\pi\sigma^2}} e^{-\frac{1}{2\sigma^2}(x_{1/2}^{(j)}-x_0)^2} \overset{!}{=} \frac{1}{2} f(x_0) = \frac{1}{2}\frac{1}{\sqrt{2\pi\sigma^2}}$$

$$\Rightarrow -\frac{1}{2\sigma^2}(x_{1/2}^{(j)} - x_0)^2 \overset{!}{=} \log(\tfrac{1}{2}) = \log(1) - \log(2) = -\log(2)$$

$$\Rightarrow (x_{1/2}^{(j)} - x_0)^2 \overset{!}{=} 2\sigma^2 \log(2) \Rightarrow x_{1/2}^{(j)} = \pm\sqrt{2\sigma^2 \log(2)} + x_0.$$

Thus, the FWHM of the normal distribution is given by

$$\mathrm{FWHM} = |x_{1/2}^{(1)} - x_{1/2}^{(2)}| = |\sqrt{2\sigma^2 \log(2)} + x_0 - (-\sqrt{2\sigma^2 \log(2)} + x_0)| = 2\sigma\sqrt{2\log(2)}.$$

To illustrate this, we consider the special case of the standard normal distribution with density $f(x) = \frac{1}{\sqrt{2\pi}} \exp(-\frac{1}{2}x^2)$. At $\pm\sqrt{2\log(2)} \approx \pm 1.18$ it is $\frac{1}{2}f_{\max} = \frac{1}{2\sqrt{2\pi}} \approx 0.199$.

The calculated values are shown in Figure 2.5.

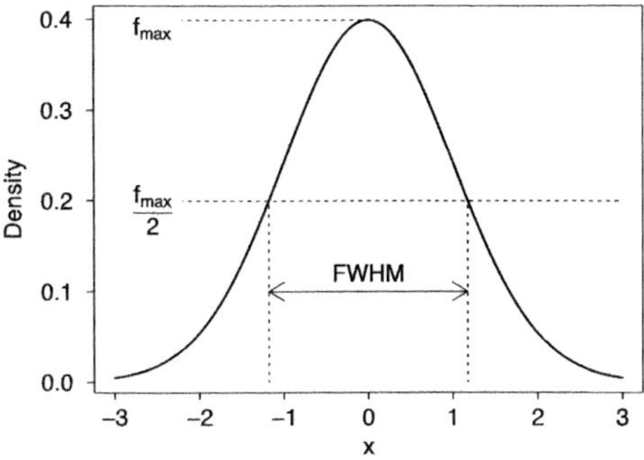

Figure 2.5 FWHM of the normal distribution with mean $x_0 = 0$ and variance $\sigma^2 = 1$

Now, we consider a Gaussian pulse $E(t) = e^{-t^2/\sigma_t^2} e^{i\omega_0 t}$ and are interested in the FWHM with respect to the time $\Delta\tau$. The maximum is at $t = 0$ and has the value $E(0) = e^0 = 1$. This way we obtain for $j \in \{1, 2\}$

$$e^{-(t_{1/2}^{(j)})^2/\sigma_t^2} \overset{!}{=} \frac{1}{2} \Rightarrow -(t_{1/2}^{(j)})^2/\sigma_t^2 \overset{!}{=} \log(\tfrac{1}{2}) = -\log(2) \Rightarrow t_{1/2}^{(j)} = \pm\sqrt{\sigma_t^2 \log(2)}$$

$$\Rightarrow \Delta\tau = |t_{1/2}^{(1)} - t_{1/2}^{(2)}| = 2\sigma_t\sqrt{\log(2)}.$$

To derive the FWHM with respect to the frequency $\Delta\omega$ the Fourier Transform of $E(t)$ is needed. First, we show that the following formula holds:

Proposition 2.3 *It is valid that*

$$\int_{-\infty}^{\infty} e^{-x^2+itx} dx = \sqrt{\pi}\, e^{-t^2/4}. \tag{2.48}$$

Proof. Using the substitution $z := x - i\frac{t}{2}$ we get

$$\int_{-\infty}^{\infty} e^{-x^2+itx} dx = e^{-t^2/4} \int_{-\infty}^{\infty} e^{-(x^2-itx-t^2/4)} dx = e^{-t^2/4} \int_{-\infty}^{\infty} e^{-(x-it/2)^2} dx$$

$$= e^{-t^2/4} \int_{-\infty}^{\infty} e^{-z^2} dz = \sqrt{\pi}\, e^{-t^2/4},$$

which corresponds exactly to the assertion. $\square$

Thus, the Fourier Transform of $E(t)$ is given by

$$
E(\omega) = \frac{1}{\sqrt{2\pi}} \int_{-\infty}^{\infty} E(t)e^{-i\omega t}\,dt
$$

$$
= \frac{1}{\sqrt{2\pi}} \int_{-\infty}^{\infty} e^{-t^2/\sigma_t^2} e^{i\omega_0 t} e^{-i\omega t}\,dt = \frac{1}{\sqrt{2\pi}} \int_{-\infty}^{\infty} e^{-s^2+i\sigma_t(\omega_0-\omega)s}\,\sigma_t\,ds
$$

$$
\overset{(2.48)}{=} \frac{1}{\sqrt{2\pi}}\sigma_t\sqrt{\pi}\,e^{-\frac{1}{4}\sigma_t^2(\omega_0-\omega)^2} = \frac{\sigma_t}{\sqrt{2}}e^{-\frac{1}{4}\sigma_t^2(\omega_0-\omega)^2},
$$

where the substitution $s := t/\sigma_t$ was used. In this case, the maximum of $E(\omega)$ is at the position $\omega = \omega_0$ and has the value $E(\omega_0) = \frac{\sigma_t}{\sqrt{2}}$.

Let $j \in \{1, 2\}$, then:

$$
e^{-\frac{1}{4}\sigma_t^2(\omega_0-\omega_{1/2}^{(j)})^2} \overset{!}{=} \frac{1}{2} \Rightarrow -\frac{1}{4}\sigma_t^2(\omega_0 - \omega_{1/2}^{(j)})^2 \overset{!}{=} -\log(2) \Rightarrow \omega_{1/2}^{(j)} = \pm\sqrt{4\log(2)/\sigma_t^2} + \omega_0
$$

$$
\Rightarrow \Delta\omega = |\omega_{1/2}^{(1)} - \omega_{1/2}^{(2)}| = 2\frac{2}{\sigma_t}\sqrt{\log(2)} = \frac{4}{\sigma_t}\sqrt{\log(2)}.
$$

Combined this leads to the relation $\Delta\tau \cdot \Delta\omega = 8\log(2)$.

2.2 Derivation of the Ponderomotive Force

Charged particles in an inhomogeneous oscillating electromagnetic field experience a nonlinear force, the so-called ponderomotive force. It was already mentioned in Chap. 1 and is derived in the following.

In [43] we consider a charged particle with mass m and charge q in the electromagnetic field of a laser pulse. Instead of being perfectly monochromatic plane waves, laser pulses have a finite width and duration. This can be described by envelope functions $\tilde{\vec{E}}$ and $\tilde{\vec{B}}$, so that the electric field $\vec{E}$ of a laser pulse with frequency ω can be written as

$$
\vec{E}(\vec{r}, t) = \mathrm{Re}\left(\tilde{\vec{E}}(\vec{r}, t)e^{-i\omega t}\right) = \mathrm{Re}\left((\mathrm{Re}(\tilde{\vec{E}}) + i\,\mathrm{Im}(\tilde{\vec{E}}))(\cos(\omega t) - i\sin(\omega t))\right)
$$

$$
= \mathrm{Re}(\tilde{\vec{E}})\cos(\omega t) + \mathrm{Im}(\tilde{\vec{E}})\sin(\omega t) = \mathrm{Re}(\tilde{\vec{E}})\frac{e^{i\omega t}+e^{-i\omega t}}{2} + \mathrm{Im}(\tilde{\vec{E}})\frac{e^{i\omega t}-e^{-i\omega t}}{2i}
$$

$$
= \tfrac{1}{2}(\mathrm{Re}(\tilde{\vec{E}}) + i\,\mathrm{Im}(\tilde{\vec{E}}))e^{-i\omega t} + \tfrac{1}{2}(\mathrm{Re}(\tilde{\vec{E}}) - i\,\mathrm{Im}(\tilde{\vec{E}}))e^{i\omega t} = \tfrac{1}{2}\tilde{\vec{E}}(\vec{r}, t)e^{-i\omega t} + \text{c.c.}
$$

and analogously the magnetic field as

$$\vec{B}(\vec{r}, t) = \mathrm{Re}\left(\tilde{\vec{B}}(\vec{r}, t)e^{-i\omega t}\right) = \tfrac{1}{2}\tilde{\vec{B}}(\vec{r}, t)e^{-i\omega t} + \mathrm{c.c.} \qquad (2.49)$$

We make the following assumptions:

- The envelope functions vary slowly, so that the time average of the electric field is

$$\left\langle \vec{E}(\vec{r}, t) \right\rangle \approx 0, \ \text{ while } \ \left\langle \tilde{\vec{E}}(\vec{r}, t) \right\rangle \neq 0. \qquad (2.50)$$

- The electron motion $\vec{r}(t)$ can be described as a superposition of a slow term $\vec{r}_s(t)$ and an oscillating term $\vec{r}_o(t)$ with $\langle \vec{r}_o(t) \rangle = 0$, so that

$$\langle \vec{r}(t) \rangle = \langle \vec{r}_s(t) + \vec{r}_o(t) \rangle = \langle \vec{r}_s(t) \rangle + \langle \vec{r}_o(t) \rangle = \langle \vec{r}_s(t) \rangle = \vec{r}_s(t). \qquad (2.51)$$

- Consider the non-relativistic regime, i.e. $\frac{v}{c} \ll 1$, and a sufficient small spatial variation of the field envelopes.

With the help of the Taylor expansion, the electric field can be split into

$$\vec{E}(\vec{r}(t), t) = \vec{E}(\vec{r}_s(t) + \vec{r}_o(t), t) \approx \vec{E}(\vec{r}_s(t), t) + (\vec{r}_o(t)^{\mathrm{T}} \nabla)\vec{E}(\vec{r}_s(t), t). \qquad (2.52)$$

To the lowest order, the equations of the oscillating components can be approximated by

$$\frac{\mathrm{d}^2 \vec{r}_o}{\mathrm{d}t^2} = \frac{\mathrm{d}\vec{v}_o}{\mathrm{d}t} \approx \frac{q}{m} \vec{E}(\vec{r}_s(t), t) \approx \frac{q}{2m} \tilde{\vec{E}}(\vec{r}_s(t))e^{-i\omega t} + \mathrm{c.c.} \qquad (2.53)$$

Proposition 2.4 *The differential equations in* (2.53) *are fulfilled by* $\vec{r}_o = \mathrm{Re}(\tilde{\vec{r}}_o e^{-i\omega t})$ *and velocity* $\vec{v}_o = \mathrm{Re}(\tilde{\vec{v}}_o e^{-i\omega t})$ *with envelopes*

$$\tilde{\vec{r}}_o = \frac{-q}{m\omega^2} \tilde{\vec{E}}(\vec{r}_s(t)) \ \text{ and } \ \tilde{\vec{v}}_o = \frac{iq}{m\omega} \tilde{\vec{E}}(\vec{r}_s(t)). \qquad (2.54)$$

Proof. The first step is to show $\frac{d^2}{dt^2}\vec{r}_o = \frac{d}{dt}\vec{v}_o$:

$$\frac{d^2\vec{r}_o}{dt^2} = \frac{d^2}{dt^2}\mathrm{Re}(\tilde{\vec{r}}_o e^{-i\omega t}) = \frac{d}{dt}\mathrm{Re}(\tilde{\vec{r}}_o(-i\omega)e^{-i\omega t}) = \frac{d}{dt}\mathrm{Re}\left(\frac{-q}{m\omega^2}\tilde{\vec{E}}(\vec{r}_s(t))(-i\omega)e^{-i\omega t}\right)$$

$$= \frac{d}{dt}\mathrm{Re}\left(\frac{iq}{m\omega}\tilde{\vec{E}}(\vec{r}_s(t))e^{-i\omega t}\right) = \frac{d}{dt}\mathrm{Re}(\tilde{\vec{v}}_o e^{-i\omega t}) = \frac{d\vec{v}_o}{dt}.$$

In the second step we prove $\frac{d}{dt}\vec{v}_o = \frac{q}{m}\vec{E}(\vec{r}_s(t),t)$:

$$\frac{d\vec{v}_o}{dt} = \frac{d}{dt}\mathrm{Re}(\tilde{\vec{v}}_o e^{-i\omega t}) = \mathrm{Re}(-i\omega\tilde{\vec{v}}_o e^{-i\omega t}) = \mathrm{Re}(-i\omega\tfrac{iq}{m\omega}\tilde{\vec{E}}(\vec{r}_s(t))e^{-i\omega t})$$

$$= \frac{q}{m}\mathrm{Re}(\tilde{\vec{E}}(\vec{r}_s(t))e^{-i\omega t})$$

$$= \frac{q}{m}\vec{E}(\vec{r}_s(t),t) = \frac{q}{m}\left(\tfrac{1}{2}\tilde{\vec{E}}(\vec{r}_s(t),t)e^{-i\omega t} + \text{c.c.}\right),$$

where the last two equalities follow from $\vec{E}(\vec{r},t) = \mathrm{Re}\left(\tilde{\vec{E}}(\vec{r},t)e^{-i\omega t}\right) = \tfrac{1}{2}\tilde{\vec{E}}(\vec{r},t)e^{-i\omega t} + \text{c.c.}$ $\square$

Proposition 2.5 *For $A(t) = \mathrm{Re}(\tilde{A}e^{-i\omega t})$ and $B(t) = \mathrm{Re}(\tilde{B}e^{-i\omega t})$ with slowly-varying $\tilde{A}$ and $\tilde{B}$, it holds*

$$\langle A(t)B(t)\rangle \approx \frac{\mathrm{Re}(\tilde{A}\tilde{B}^*)}{2}. \tag{2.55}$$

Proof. The splitting of the envelopes into their real and imaginary parts leads to

$$\langle A(t)B(t)\rangle = \left\langle\mathrm{Re}(\tilde{A}e^{-i\omega t})\mathrm{Re}(\tilde{B}e^{-i\omega t})\right\rangle$$

$$= \Big\langle\mathrm{Re}\left((\mathrm{Re}(\tilde{A}) + i\mathrm{Im}(\tilde{A}))(\cos(\omega t) - i\sin(\omega t))\right)\cdot$$

$$\cdot\,\mathrm{Re}\left((\mathrm{Re}(\tilde{B}) + i\mathrm{Im}(\tilde{B}))(\cos(\omega t) - i\sin(\omega t))\right)\Big\rangle$$

$$= \Big\langle(\mathrm{Re}(\tilde{A})\cos(\omega t) + \mathrm{Im}(\tilde{A})\sin(\omega t))(\mathrm{Re}(\tilde{B})\cos(\omega t) + \mathrm{Im}(\tilde{B})\sin(\omega t))\Big\rangle$$

$$= \Big\langle\mathrm{Re}(\tilde{A})\mathrm{Re}(\tilde{B})\cos^2(\omega t) + (\mathrm{Re}(\tilde{A})\mathrm{Im}(\tilde{B}) + \mathrm{Im}(\tilde{A})\mathrm{Re}(\tilde{B}))\sin(\omega t)\cos(\omega t) +$$

$$+ \mathrm{Im}(\tilde{A})\mathrm{Im}(\tilde{B})\sin^2(\omega t)\Big\rangle$$

$$\approx \mathrm{Re}(\tilde{A})\mathrm{Re}(\tilde{B})\big\langle\cos^2(\omega t)\big\rangle + (\mathrm{Re}(\tilde{A})\mathrm{Im}(\tilde{B}) + \mathrm{Im}(\tilde{A})\mathrm{Re}(\tilde{B}))\,\langle\sin(\omega t)\cos(\omega t)\rangle +$$

$$+ \mathrm{Im}(\tilde{A})\mathrm{Im}(\tilde{B})\big\langle\sin^2(\omega t)\big\rangle$$

$$= \tfrac{1}{2}\left(\mathrm{Re}(\tilde{A})\mathrm{Re}(\tilde{B}) + \mathrm{Im}(\tilde{A})\mathrm{Im}(\tilde{B})\right) = \tfrac{1}{2}\mathrm{Re}\left((\mathrm{Re}(\tilde{A}) + i\mathrm{Im}(\tilde{A}))(\mathrm{Re}(\tilde{B}) - i\mathrm{Im}(\tilde{B}))\right)$$

$$= \tfrac{1}{2}\mathrm{Re}\left(\tilde{A}\tilde{B}^*\right).$$

where the relations $\langle \sin(\omega t)\cos(\omega t)\rangle = 0$ and $\langle \sin^2(\omega t)\rangle = \langle \cos^2(\omega t)\rangle = \frac{1}{2}$ were used. $\qquad\square$

Now, the time average of the electric field can be written as

$$\langle \vec{E}(\vec{r}(t),t)\rangle \overset{(2.52)}{\approx} \langle \vec{E}(\vec{r}_s(t),t) + (\vec{r}_o(t)^{\mathrm{T}}\nabla)\vec{E}(\vec{r}_s(t),t)\rangle \overset{(2.55)}{\approx} \tfrac{1}{4}(\tilde{\vec{r}}_o^{\,*}(t)^{\mathrm{T}}\nabla)\tilde{\vec{E}}(\vec{r}_s(t),t) + \text{c.c.}$$

$$\overset{(2.54)}{=} \frac{-q}{4m\omega^2}(\tilde{\vec{E}}^{\,*}(\vec{r}_s(t),t)^{\mathrm{T}}\nabla)\tilde{\vec{E}}(\vec{r}_s(t),t) + \text{c.c.}$$

Using the third Maxwell equation $\nabla \times \vec{E} = -\partial_t \vec{B} = i\omega \vec{B}$, the time average of the cross product of $\vec{v}$ and $\vec{B}$ can be approximated by

$$\langle \vec{v} \times \vec{B}(\vec{r}(t),t)\rangle \overset{(2.55)}{\approx} \frac{1}{4}\tilde{\vec{v}}_o^{\,*} \times \tilde{\vec{B}}(\vec{r}_s(t),t) + \text{c.c.}$$

$$\overset{(2.54)}{=} \frac{1}{4}\frac{-iq}{m\omega}\tilde{\vec{E}}^{\,*}(\vec{r}_s(t),t) \times \left(\tfrac{1}{i\omega}\nabla \times \tilde{\vec{E}}(\vec{r}_s(t),t)\right) + \text{c.c.}$$

$$= \frac{-q}{4m\omega^2}\tilde{\vec{E}}^{\,*}(\vec{r}_s(t),t) \times \left(\nabla \times \tilde{\vec{E}}(\vec{r}_s(t),t)\right) + \text{c.c.}$$

Furthermore, the spatial derivative of a scalar product will be needed. For this purpose the following relation that can be found in [44] will be used:

Proposition 2.6 *For two three-dimensional vectors $\vec{A} := \vec{A}(\vec{r})$ and $\vec{B} := \vec{B}(\vec{r})$, the spatial derivative $\nabla := \frac{\partial}{\partial \vec{r}}$ of the scalar product of these vectors can be written as*

$$\nabla(\vec{A}^{\mathrm{T}}\vec{B}) = (\vec{A}^{\mathrm{T}}\nabla)\vec{B} + (\vec{B}^{\mathrm{T}}\nabla)\vec{A} + \vec{A} \times (\nabla \times \vec{B}) + \vec{B} \times (\nabla \times \vec{A}). \tag{2.56}$$

Proof. For $i \in \{1,2,3\}$ and with the Levi-Civita symbol ε_{ijk} and the Kronecker delta $\delta_{i,j}$ we are able to determine the i-th component of $\vec{A} \times (\nabla \times \vec{B}) + \vec{B} \times (\nabla \times \vec{A})$ as

$$(\vec{A} \times (\nabla \times \vec{B}) + \vec{B} \times (\nabla \times \vec{A}))_i = \varepsilon_{ijk}(A_j(\nabla \times \vec{B})_k + B_j(\nabla \times \vec{A})_k)$$

$$= \varepsilon_{ijk}A_j\varepsilon_{klm}\partial_l B_m + \varepsilon_{ijk}B_j\varepsilon_{klm}\partial_l A_m$$

$$= \varepsilon_{ijk}\varepsilon_{klm}(A_j\partial_l B_m + B_j\partial_l A_m)$$

$$= (\delta_{i,l}\delta_{j,m} - \delta_{i,m}\delta_{j,l})(A_j\partial_l B_m + B_j\partial_l A_m)$$

$$= A_j\partial_i B_j + B_j\partial_i A_j - A_j\partial_j B_i - B_j\partial_j A_i$$

$$= (\nabla(\vec{A}^{\mathrm{T}}\vec{B}) - (\vec{A}^{\mathrm{T}}\nabla)\vec{B} - (\vec{B}^{\mathrm{T}}\nabla)\vec{A})_i,$$

where the Einstein summation convention was used. $\qquad\square$

In particular, this leads to the corollary

$$
\begin{aligned}
\nabla |\vec{A}|^2 = \nabla(\vec{A}^{\mathrm{T}}\vec{A}^*) &\overset{(2.56)}{=} (\vec{A}^{\mathrm{T}}\nabla)\vec{A}^* + ((\vec{A}^*)^{\mathrm{T}}\nabla)\vec{A} + \vec{A}\times(\nabla\times\vec{A}^*) + \vec{A}^*\times(\nabla\times\vec{A}) \\
&= ((\vec{A}^*)^{\mathrm{T}}\nabla)\vec{A} + \vec{A}^*\times(\nabla\times\vec{A}) + \text{c.c.}
\end{aligned}
\tag{2.57}
$$

Additionally, we get with the assumption of a slowly varying envelope function:

$$
\begin{aligned}
\left\langle \vec{E}^2(\vec{r}_s, t) \right\rangle &= \left\langle \left(\tfrac{1}{2}\tilde{\vec{E}}(\vec{r}_s, t)e^{-i\omega t} + \tfrac{1}{2}\tilde{\vec{E}}^*(\vec{r}_s, t)e^{i\omega t} \right)^2 \right\rangle \\
&= \left\langle \tfrac{1}{4}\tilde{\vec{E}}^2(\vec{r}_s, t)e^{-2i\omega t} + \tfrac{2}{4}\tilde{\vec{E}}^{\mathrm{T}}(\vec{r}_s, t)\tilde{\vec{E}}^*(\vec{r}_s, t) + \tfrac{1}{4}(\tilde{\vec{E}}^*)^2(\vec{r}_s, t)e^{2i\omega t} \right\rangle \\
&\approx \tfrac{1}{2}\tilde{\vec{E}}^{\mathrm{T}}(\vec{r}_s, t)\tilde{\vec{E}}^*(\vec{r}_s, t) = \tfrac{1}{2}|\tilde{\vec{E}}(\vec{r}_s, t)|^2.
\end{aligned}
\tag{2.58}
$$

Averaging the Lorentz force we obtain

$$
\begin{aligned}
m\frac{\mathrm{d}\vec{v}_s}{\mathrm{d}t} = m\frac{\mathrm{d}^2\vec{r}_s}{\mathrm{d}t^2} = m\frac{\mathrm{d}^2\langle\vec{r}\rangle}{\mathrm{d}t^2} = m\frac{\mathrm{d}\langle\vec{v}\rangle}{\mathrm{d}t} &= q\left\langle \vec{E}(\vec{r}(t),t) \right\rangle + q\left\langle \vec{v}\times\vec{B}(\vec{r}(t),t) \right\rangle \\
&\approx \frac{-q^2}{4m\omega^2}(\tilde{\vec{E}}^*(\vec{r}_s(t),t)^{\mathrm{T}}\nabla)\tilde{\vec{E}}(\vec{r}_s(t),t) - \frac{q^2}{4m\omega^2}\tilde{\vec{E}}^*(\vec{r}_s(t),t)\times\left(\nabla\times\tilde{\vec{E}}(\vec{r}_s(t),t)\right) + \text{c.c.} \\
&= \frac{-q^2}{4m\omega^2}\left((\tilde{\vec{E}}^*(\vec{r}_s(t),t)^{\mathrm{T}}\nabla)\tilde{\vec{E}}(\vec{r}_s(t),t) + \tilde{\vec{E}}^*(\vec{r}_s(t),t)\times\left(\nabla\times\tilde{\vec{E}}(\vec{r}_s(t),t)\right) \right) + \text{c.c.} \\
&\overset{(2.57)}{=} \frac{-q^2}{4m\omega^2}\nabla\left|\tilde{\vec{E}}(\vec{r}_s(t),t)\right|^2 \overset{(2.58)}{\approx} \frac{-q^2}{2m\omega^2}\nabla\left\langle \vec{E}^2(\vec{r}_s(t),t) \right\rangle =: \vec{F}_p,
\end{aligned}
$$

what we designate as **ponderomotive force** $\vec{F}_p$. The ponderomotive force has very interesting characteristics:

- It only contains q^2, which means that the sign of the charge does not matter. For instance, an electron and a positron experience the same ponderomotive force.
- Due to the proportionality of $\vec{F}_p$ to $\frac{1}{m}$, the ponderomotive force of ions can be neglected with respect to that of free electrons.
- The ponderomotive force is proportional to $-\nabla\left\langle \vec{E}^2 \right\rangle$. This implies that $\vec{F}_p$ acts in the direction of the steepest descent of $\left\langle \vec{E}^2 \right\rangle$, so particles tend to move in regions, where the field is weaker.

In the relativistic regime, the ponderomotive force is of a similar form. Essentially, the two formulas differ by a factor γ. A detailed derivation can be found in [45].

Remark: Note that there are two typos in [43] on page 16: In Equation (2.45) there should be "+" instead of "$-$", because in the slightly different notation used there it holds

$$\tilde{v_0}^{*} \overset{(2.42)}{=} \left(-\frac{ie}{m_e\omega}\tilde{\mathbf{E}}(\mathbf{r}_s(t)) \right)^{*} = \frac{+ie}{m_e\omega}\tilde{\mathbf{E}}^{*}(\mathbf{r}_s(t)).$$

This typo led to another typo, which is in Equation (2.46). There it is used

$$\left(\tilde{\mathbf{E}}^{*}(\mathbf{r}_s(t),t)\cdot\nabla \right)\tilde{\mathbf{E}}(\mathbf{r}_s(t),t) - \tilde{\mathbf{E}}^{*}(\mathbf{r}_s(t),t)\times(\nabla\times\tilde{\mathbf{E}}(\mathbf{r}_s(t),t)) + \text{c.c.} = \nabla|\tilde{\mathbf{E}}(\mathbf{r}_s(t),t)|^2,$$

but as derived above, Corollary (2.57) holds and thus there needs to be a plus sign instead of a minus sign. Both sign errors cancel each other out and that is why the final formula remains correct.

2.3 Relativistic Lagrangian

In this subchapter, we study the relativistic motion of a single charged particle in an electromagnetic field. Firstly, the equations of motion are derived as generally as possible and then calculated for a concrete example.

2.3.1 Equation of Motion

Consider a particle with rest mass m, velocity $\vec{v}$ and charge q in an electromagnetic field with scalar potential ϕ and vector potential $\vec{A}$. Following [46], in special relativity, the Lagrangian for this particle is given by

$$L = -mc^2\sqrt{1 - \frac{v^2}{c^2}} - q\phi + q\vec{v}^{\mathsf{T}}\vec{A}. \tag{2.59}$$

Before we derive the equations of motion using the Lagrangian formalism, we first consider the Lorentz factor $\gamma = 1/\sqrt{1 - \frac{v^2}{c^2}}$ in more detail. With the relativistic momentum $p = \gamma m v$ and the limits $\gamma(v = 0) = 1$ and $\lim_{|v|\to c}\gamma(v) = \infty$ it follows

$$\gamma = \frac{1}{\sqrt{1-\frac{v^2}{c^2}}} \overset{\gamma>0}{\Leftrightarrow} \gamma^2 = \frac{1}{1-\frac{v^2}{c^2}} \overset{v\le c}{\Leftrightarrow} \gamma^2\left(1-\underbrace{\left(\frac{v}{c}\right)^2}_{\in[0,1[}\right) = 1 \Leftrightarrow \gamma^2 = 1 + \gamma^2\underbrace{\left(\frac{v}{c}\right)^2}_{}$$

$$\underbrace{}_{\in\,]0,1]}$$

$$\Leftrightarrow \gamma^2 = 1 + \left(\frac{\gamma m v}{mc}\right)^2 \Leftrightarrow \gamma^2 = 1 + \left(\frac{p}{mc}\right)^2 \overset{\gamma>0}{\Leftrightarrow} \gamma = \sqrt{1 + \left(\frac{p}{mc}\right)^2}.$$

$$(2.60)$$

Another common way to express the Lorentz factor is by splitting the total energy into a potential part and a kinetic part, i.e. $E := E_{\text{total}} = E_0 + E_{\text{kin}}$, and by using the energy-momentum relation $E^2 = p^2c^2 + m^2c^4$. This leads to

$$\frac{E_{\text{kin}}}{E_0} + 1 = \frac{E-E_0}{E_0} + 1 = \sqrt{\frac{E^2}{E_0^2} \underbrace{- \frac{E_0}{E_0} + 1}_{=0}} = \sqrt{\frac{p^2c^2+m^2c^4}{m^2c^4}} = \sqrt{\frac{p^2}{m^2c^2} + 1} = \sqrt{1 + \left(\frac{p}{mc}\right)^2} \overset{(2.60)}{=} \gamma.$$

$$(2.61)$$

In addition, the total derivative of the potential $\vec{A} := \vec{A}(\vec{r}, t)$ with respect to time is needed.

Proposition 2.7 *The total derivative of the vector potential $\vec{A}$ is given by*

$$\frac{d}{dt}\vec{A} = \frac{\partial}{\partial t}\vec{A} + \left(\vec{v}^T\nabla\right)\vec{A}. \tag{2.62}$$

Proof. Using the chain rule we get for the total derivative of the i-th component of $\vec{A}$

$$\frac{d}{dt}A_i = \frac{\partial}{\partial t}A_i + \frac{\partial A_i}{\partial x}\underbrace{\frac{dx}{dt}}_{=v_x} + \frac{\partial A_i}{\partial y}\underbrace{\frac{dy}{dt}}_{=v_y} + \frac{\partial A_i}{\partial z}\underbrace{\frac{dz}{dt}}_{=v_z}$$

$$= \frac{\partial}{\partial t}A_i + (v_x\partial_x + v_y\partial_y + v_z\partial_z)A_i = \frac{\partial}{\partial t}A_i + (\vec{v}^T\nabla)A_i,$$

where $i \in \{1, 2, 3\}$. All in all, this results in the expression $\frac{d}{dt}\vec{A} = \frac{\partial}{\partial t}\vec{A} + (\vec{v}^T\nabla)\vec{A}$. $\qquad\square$

After our careful preparation we can now derive the equations of motion using the Lagrangian formalism. The first step is the calculation of the canonical momentum

$\vec{p}_c$, which results from the partial derivative of the Lagrangian (2.59) with respect to the velocity:

$$\vec{p}_c := \frac{\partial L}{\partial \vec{v}} = -mc^2 \cdot \frac{1}{2}\gamma \frac{-2\vec{v}}{c^2} + q\vec{A} = \gamma m\vec{v} + q\vec{A} = \vec{p} + q\vec{A}.$$

With $v^2 = \vec{v}^{\mathsf{T}}\vec{v}$ we get for the Hamiltonian

$$H := \vec{p}_c^{\mathsf{T}}\vec{v} - L = (\vec{p} + q\vec{A})^{\mathsf{T}}\vec{v} + mc^2\sqrt{1 - \frac{v^2}{c^2}} + q\phi - q\vec{v}^{\mathsf{T}}\vec{A} = \gamma mv^2 + mc^2\sqrt{1 - \frac{v^2}{c^2}} + q\phi$$

$$= \gamma mc^2\left(\frac{v^2}{c^2} + \frac{1}{\gamma^2}\right) + q\phi = \gamma mc^2\left(\frac{v^2}{c^2} + 1 - \frac{v^2}{c^2}\right) + q\phi = \gamma mc^2 + q\phi$$

$$= \sqrt{p^2c^2 + m^2c^4} + q\phi = c\sqrt{p^2 + m^2c^2} + q\phi = c\sqrt{(p_c - qA)^2 + m^2c^2} + q\phi.$$

We recall that the electric field $\vec{E}$ and the magnetic field $\vec{B}$ can be represented as $\vec{B} = \nabla \times \vec{A}$ and $\vec{E} = -\nabla\phi - \frac{\partial}{\partial t}\vec{A}$. This can be used to show that the solution of the Euler-Lagrange equation

$$\frac{\partial L}{\partial \vec{r}} - \frac{\mathrm{d}}{\mathrm{d}t}\frac{\partial L}{\partial \vec{v}} = -q\nabla\phi + q\nabla(\vec{v}^{\mathsf{T}}\vec{A}) - \frac{\mathrm{d}}{\mathrm{d}t}(\vec{p} + q\vec{A})$$

$$\overset{(2.56)}{=} -q\nabla\phi + q(\underbrace{(\vec{v}^{\mathsf{T}}\nabla)\vec{A} + (\vec{A}^{\mathsf{T}}\nabla)\vec{v}}_{=0} + \vec{v} \times (\nabla \times \vec{A}) + \underbrace{\vec{A} \times (\nabla \times \vec{v})}_{=0}) - \frac{\mathrm{d}}{\mathrm{d}t}\vec{p} - q\frac{\mathrm{d}}{\mathrm{d}t}\vec{A}$$

$$\overset{(2.62)}{=} -q\nabla\phi + q(\vec{v}^{\mathsf{T}}\nabla)\vec{A} + q\vec{v} \times (\nabla \times \vec{A}) - \frac{\mathrm{d}}{\mathrm{d}t}\vec{p} - q(\frac{\partial}{\partial t}\vec{A} + (\vec{v}^{\mathsf{T}}\nabla)\vec{A})$$

$$= q\underbrace{(-\nabla\phi - \frac{\partial}{\partial t}\vec{A})}_{=\vec{E}} + q(\vec{v}^{\mathsf{T}}\nabla)\vec{A} + q\vec{v} \times \underbrace{(\nabla \times \vec{A})}_{=\vec{B}} - \frac{\mathrm{d}}{\mathrm{d}t}\vec{p} - q(\vec{v}^{\mathsf{T}}\nabla)\vec{A}$$

$$= q\vec{E} + q\vec{v} \times \vec{B} - \frac{\mathrm{d}}{\mathrm{d}t}\vec{p} \overset{!}{=} 0$$

is given by the Lorentz force $\vec{F}_{\mathrm{L}} := \frac{\mathrm{d}}{\mathrm{d}t}\vec{p} = q\vec{E} + q\vec{v} \times \vec{B}$. Additionally, we can use the relativistic momentum $\vec{p} = \gamma m\vec{v}$ to get $\frac{\mathrm{d}}{\mathrm{d}t}\vec{r} = \vec{v} = \frac{\vec{p}}{\gamma m}$.

2.3.2 Example

Consider a plane wave with vector potential

$$\vec{A}(x, t) = \vec{e}_y\left[\frac{1}{2}A_0\left(e^{i(\omega_0 t - k_0 x)} + \text{c.c.}\right) + \frac{1}{2}A_1\left(e^{i(\omega_1 t - k_1 x)} + \text{c.c}\right)\right].$$

The assumptions $\omega_1 = \omega_0$ and $k_1 = -k_0$ lead to the simplification

$$\vec{A}(x, t) = \vec{e}_y [A_0 \cos(\omega_0 t - k_0 x) + A_1 \cos(\omega_0 t + k_0 x)],$$

using the representation of the cosine: $\cos(x) = \frac{1}{2}\left(e^{ix} + e^{-ix}\right)$. The envelopes can be written as $A_0(x, t) = \frac{1}{2}a_0(1 - \tanh(\frac{x-t}{\sigma}))$ and $A_1(x, t) = \frac{1}{2}a_1(1 - \tanh(\frac{-x-t}{\sigma}))$. In order to determine the concrete equations of motion, the electric and the magnetic field are calculated.

Because of $\phi = 0$ and $A_x = A_z = 0$ the electric field is of the shape

$$E(x, t) = -\nabla\phi - \partial_t \vec{A} = \begin{pmatrix} 0 \\ -\partial_t A_y \\ 0 \end{pmatrix} = \begin{pmatrix} 0 \\ E_y \\ 0 \end{pmatrix}.$$

With the derivative of the hyperbolic tangent $\frac{\mathrm{d}}{\mathrm{d}t}\tanh(t) = 1/\cosh^2(t)$ the derivatives of the envelopes with respect to time and space can be calculated:

$$\partial_t A_{0/1} = \frac{a_{0/1}}{2} \cdot (\mp 1)\frac{1}{\cosh^2(\frac{\pm x - t}{\sigma})} \cdot \frac{\mp 1}{\sigma} = \frac{a_{0/1}}{2} \cdot \frac{1}{\cosh^2(\frac{\pm x - t}{\sigma})} \cdot \frac{1}{\sigma},$$

$$\partial_x A_{0/1} = \frac{1}{2}a_{0/1}(-1)\frac{1}{\cosh^2(\frac{\pm x - t}{\sigma})}\frac{\pm 1}{\sigma} = \mp\frac{a_{0/1}}{2\sigma}\frac{1}{\cosh^2(\frac{\pm x - t}{\sigma})}.$$

This leads to the y-component of the electric field

$$\begin{aligned} E_y = -\partial_t A_y &= A_0\omega_0 \sin(\omega_0 t - k_0 x) + A_1\omega_0 \sin(\omega_0 t + k_0 x) \\ &\quad - \frac{\partial A_0}{\partial t}\cos(\omega_0 t - k_0 x) - \frac{\partial A_1}{\partial t}\cos(\omega_0 t + k_0 x). \end{aligned}$$

Let us now turn to the magnetic field:

$$\vec{B}(x, t) = \nabla \times \vec{A} = \begin{pmatrix} \partial_y A_z - \partial_z A_y \\ \partial_z A_x - \partial_x A_z \\ \partial_x A_y - \partial_y A_x \end{pmatrix} = \begin{pmatrix} -\partial_z A_y \\ 0 \\ \partial_x A_y \end{pmatrix} = \begin{pmatrix} 0 \\ 0 \\ \partial_x A_y \end{pmatrix} = \begin{pmatrix} 0 \\ 0 \\ B_z \end{pmatrix} \quad \text{with}$$

$$\begin{aligned} B_z = \partial_x A_y &= A_0 k_0 \sin(\omega_0 t - k_0 x) - A_1 k_0 \sin(\omega_0 t + k_0 x) \\ &\quad + \frac{\partial A_0}{\partial x}\cos(\omega_0 t - k_0 x) + \frac{\partial A_1}{\partial x}\cos(\omega_0 t + k_0 x). \end{aligned}$$

Finally, we are able to state the equations of motion

$$\dot{\vec{p}} = -(\vec{E} + \vec{v} \times \vec{B}) = -\left(\vec{E} + \begin{pmatrix} v_y B_z - v_z B_y \\ v_z B_x - v_x B_z \\ v_x B_y - v_y B_x \end{pmatrix} \right)$$

$$= -\begin{pmatrix} 0 \\ E_y \\ 0 \end{pmatrix} - \begin{pmatrix} v_y B_z \\ -v_x B_z \\ 0 \end{pmatrix} = \begin{pmatrix} -v_y B_z \\ v_x B_z - E_y \\ 0 \end{pmatrix},$$

$$\dot{\vec{x}} = \frac{\vec{p}}{\gamma}.$$

Note that the natural units are used as described previously.

Wakefield Acceleration

This chapter begins with a description of laser-driven plasma wakefield acceleration and rough estimates for the size of the plasma cavity and the electric field with which the electrons are accelerated. We are interested in the arrangement of electrons in the accelerated electron bunch. The assumptions that lead to the Hamiltonian, which describes the energy of this system, are then explained. We follow the derivation of the Hamiltonian in [16]. Since we are looking for those particle positions for which the Hamiltonian becomes minimal, two approaches to minimize the Hamiltonian numerically are discussed. One of these approaches, the Gradient Descent Method, is examined in more detail in the next chapter. In preparation for this analysis, we will reduce the Hamiltonian to its essential properties.

3.1 General Setting

A high-intensity laser pulse penetrates a homogeneous electron-ion plasma. Due to the electromagnetic field of the laser pulse the electrons close to it experience a ponderomotive force which displaces them from the center (see Fig. 3.1) predominantly perpendicular to the direction of propagation.

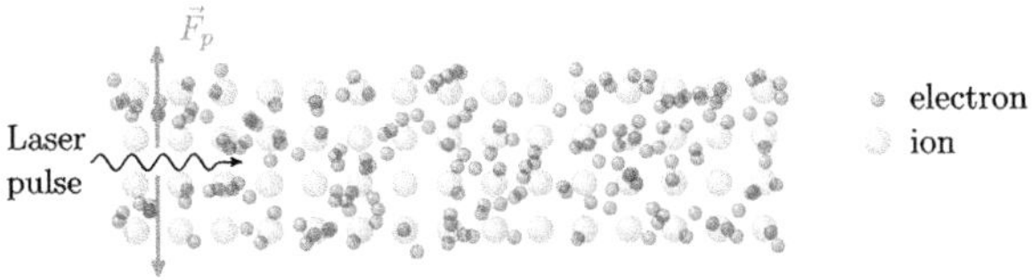

Figure 3.1 Displacement of electrons by the laser pulse due to the ponderomotive force $\vec{F}_p$

M. Hagedorn, *Minimization Problems for the Witness Beam in Relativistic Plasma Cavities*, BestMasters, https://doi.org/10.1007/978-3-658-46226-0_3

As shown in Sect. 2.2, the ponderomotive force is anti-proportional to the mass and therefore negligible for ions on the considered time scale. The remaining positive region directly behind the laser pulse re-attracts the electrons. This restoring force causes that the positive region is followed by a negative region and that is how a small area with a very strong field in longitudinal direction is created. We call this region bubble or plasma cavity. It can be used to accelerate electrons to almost the speed of light. In addition, the transverse field components have a focusing effect. This principle is called wakefield acceleration and was proposed for the first time in [47].

As the electrons driven by the restoring force can overshoot, this results in a periodic plasma density perturbation and is illustrated in Fig. 3.2.

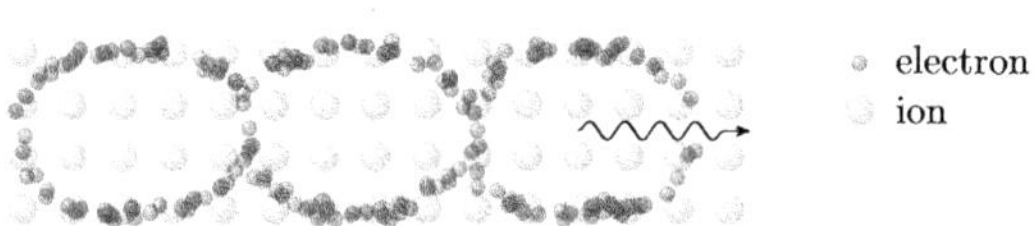

Figure 3.2 Wakefield driven by the laser pulse

Analogously, it is also possible to use electron [48] or proton [49] beams instead of a laser pulse for the preparation of the plasma.

To estimate the size of the plasma cavity, i. e. the positive region directly behind the laser pulse, we assume that the cavity moves with the speed of light c and that the electrons oscillate with the plasma frequency ω_p (see (2.7)). Thus, the wavelength of the oscillation corresponds to the distance d between the laser pulse and the negative region behind the plasma cavity. Using the relations $f = \frac{c}{\lambda}$ and $\omega = 2\pi f$, the diameter d of the cavity in longitudinal direction can be estimated by

$$d = \lambda = \frac{c}{f} = \frac{c}{\omega_p/(2\pi)} = 2\pi c \sqrt{\frac{\varepsilon_0 m_e}{e^2 n_e}}. \tag{3.1}$$

This corresponds to twice the radius if a spherical bubble is assumed. Concretely, a critical density of $n_c = 2 \cdot 10^{21} \frac{1}{\text{cm}^3}$ and an electron density of

$$n_e = 0.5\% \cdot n_c = 0.5 \cdot 10^{-2} \cdot 2 \cdot 10^{21} \frac{1}{\text{cm}^3} = 10^{19} \frac{1}{\text{cm}^3} = 10^{25} \frac{1}{\text{m}^3}, \tag{3.2}$$

as assumed in [11], leads to a cavity size of approximately

$$d = 2\pi c \sqrt{\frac{\varepsilon_0 m_e}{e^2 n_e}} \approx 2\pi \cdot 2.998 \cdot 10^8 \tfrac{\text{m}}{\text{s}} \cdot \sqrt{\frac{8.854 \cdot 10^{-12} \tfrac{\text{As}}{\text{Vm}} \cdot 9.109 \cdot 10^{-31} \text{kg}}{(1.602 \cdot 10^{-19} \text{As})^2 \cdot 10^{25} \tfrac{1}{\text{m}^3}}}$$

$$\approx 1 \cdot 10^{-5} \tfrac{\text{m}}{\text{s}} \sqrt{\frac{\text{As}^3}{\text{kg} \cdot \text{m}^2} \frac{\text{As}}{(\text{As})^2} \frac{\text{kg} \cdot \text{m}^3}{\text{m}}} = 10 \cdot 10^{-6} \tfrac{\text{m}}{\text{s}} \sqrt{\text{s}^2} = 10 \, \mu\text{m}.$$

There are several methods of placing electrons inside the cavity, so that they can be accelerated. Under certain conditions it is possible to trap electrons directly as discussed in [50]. Other possibilities, for instance, are the ionization-induced electron injection [12] or using a density down-ramp [51]. However, in the following we assume the situation of having an electron bunch inside the cavity (like in Fig. 3.3) and are interested in analyzing the arrangement of these electrons. In this context, the electron bunch is also referred to as a witness beam.

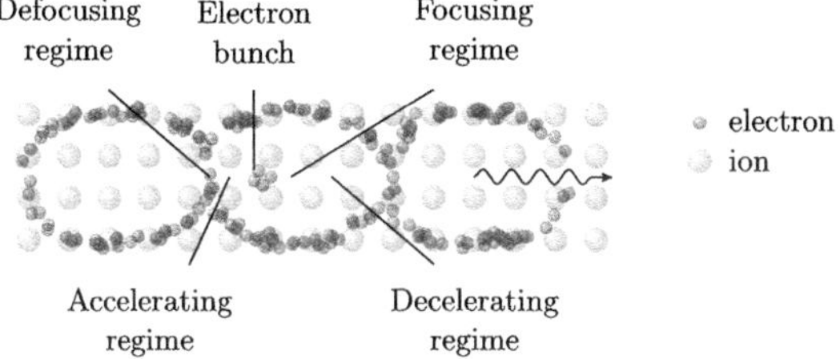

Figure 3.3 Different regimes when accelerating an electron bunch inside the bubble

The developed sketch, Fig. 3.3, is based on the drawing in [52], which, however, refers to a particle-driven plasma wakefield acceleration.

In order to get an intuition for the magnitude of such electric fields generated by the ponderomotive force, we follow the derivation of an upper bound of the strength of the accelerating electric field, which is given in Chapter 6 of [21]. The maximum electric field is obtained if the inverse of the pulse duration τ is equal to the eigenfrequency of the plasma $f_{\text{eigen}} = \frac{\omega_p}{2\pi}$. Assuming that all electrons are displaced from the area of the laser pulse and approximating the left hand side of the Maxwell equation

$$\nabla \cdot \vec{E} = \frac{\rho}{\varepsilon_0} = \frac{-e\, n_e}{\varepsilon_0} \tag{3.3}$$

by $\nabla \cdot \vec{E} \approx k_p E_{\text{max}}$ with wavenumber k_p of the plasma, we get:

$$E_{\text{max}} \approx \frac{e\,n_e}{\varepsilon_0}\frac{1}{k_p} = \frac{e\,n_e}{\varepsilon_0}\frac{\lambda}{2\pi} = \frac{e\,n_e}{\varepsilon_0}\frac{c}{2\pi\,f_{\text{eigen}}} = \frac{e\,n_e}{\varepsilon_0}\frac{c}{\omega_p} = \frac{e\,c\,n_e}{\varepsilon_0}\sqrt{\frac{\varepsilon_0\,m_e}{e^2 n_e}} = \sqrt{\frac{m_e c^2 n_e}{\varepsilon_0}}.$$

$$(3.4)$$

For the above used electron density of $n_e = 10^{25}\,\frac{1}{\text{m}^3}$ this corresponds to

$$E_{\text{max}} \approx \sqrt{\frac{m_e c^2 n_e}{\varepsilon_0}} \approx \sqrt{\frac{9.109 \cdot 10^{-31}\text{kg} \cdot (2.998 \cdot 10^8\,\frac{\text{m}}{\text{s}})^2 \cdot 10^{25}\,\frac{1}{\text{m}^3}}{8.854 \cdot 10^{-12}\,\frac{\text{A s}}{\text{V m}}}} \approx 3 \cdot 10^{12}\,\frac{\text{V}}{\text{m}} = 3\,\frac{\text{TV}}{\text{m}}.$$

$$(3.5)$$

3.2 Description of the System Using the Hamiltonian

The aim of this subchapter is to derive and discuss the Hamiltonian that describes the energy of the electrons in the witness beam in the equilibrium state. It is based on the derivation in [16].

We consider the movement of the bubble as a cylindrically symmetrical system with propagation direction z. Using the natural units from Sect. 2.1.7 the bubble velocity $V_0 \approx c$ can be expressed by the initial γ-factor γ_0 as follows:

$$V_0 = \sqrt{1 - \frac{1}{\gamma_0^2}}.$$

$$(3.6)$$

Without loss of generality we choose a co-moving reference frame by using the canonical transformation

$$\xi = z - V_0 t.$$

$$(3.7)$$

Thus, ξ is nothing else than the longitudinal position of the particle inside the bubble in the moving system. We split the occurring fields into an external component and an interaction component. We use a strongly simplified quasi-static approximation [10] that allows us to assume the external potentials to be known. In this model, the acceleration in the z-direction is determined solely by the wakefield potential

$$\Psi(\vec{r}_i) = \tfrac{1}{8}(x_i^2 + y_i^2 + \xi_i^2).$$

$$(3.8)$$

The electromagnetic interaction of resting particles could be described by the Coulomb potential. However, in this situation it is assumed that the particles move with high velocities so that the Coulomb approximation is not appropriate. It needs to be considered that the time t of the observation at r of a particle at $\tilde{r}$ is different to the retarded time $t_{\text{ret}} := t - \frac{1}{c}|r - \tilde{r}|$ when the observed particle actually was at $\tilde{r}$. The retarded time needs to be considered because the electromagnetic force acting between the two particles is mediated by a photon moving with finite velocity c. In other words, an event B at space-time (t_B, r_B) can only be caused by an event A at (t_A, r_A) with $t_A < t_B$, if the space-time distance

$$\mathrm{d}s = \sqrt{c^2(\mathrm{d}t)^2 - (\mathrm{d}x)^2 - (\mathrm{d}y)^2 - (\mathrm{d}z)^2} \qquad (3.9)$$

is real, i.e. if $t_B - t_A \geq \|r_B - r_A\|/c$ [53]. This is visualized in Figure 3.4.

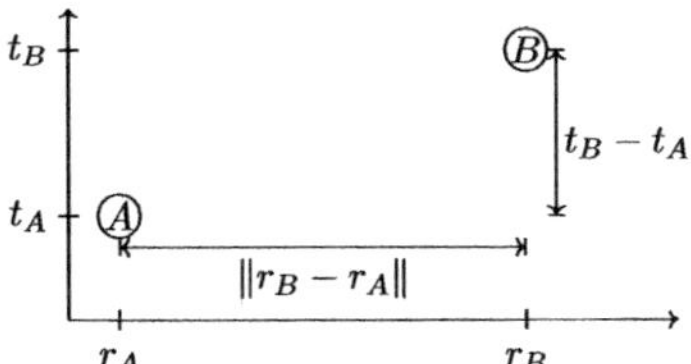

Figure 3.4 Events A and B in space-time

Therefore, we describe the electron-electron interaction by the retarded Liénard-Wiechert potentials. For a point-like particle i at position $\vec{r}_i$ and retarded time t_{ret} with velocity $\vec{v}_i$ and charge q_i measured at position $\vec{r}$ and time t the scalar potential φ and the vector potential $\vec{A}$ are

$$
\begin{aligned}
\varphi &= \sum_{i=1}^{n} \frac{q_i}{|\vec{r}(t) - \vec{r}_i(t_{\text{ret}})| - \frac{\vec{v}_i(t_{\text{ret}})}{c}[\vec{r}(t) - \vec{r}_i(t_{\text{ret}})]}, \\
\vec{A} &= \sum_{i=1}^{n} \frac{q_i \vec{v}_i(t_{\text{ret}})}{|\vec{r}(t) - \vec{r}_i(t_{\text{ret}})| - \frac{\vec{v}_i(t_{\text{ret}})}{c}[\vec{r}(t) - \vec{r}_i(t_{\text{ret}})]}.
\end{aligned}
\qquad (3.10)
$$

Thus, additionally to the terms for a free particle and the coupling to external potentials, which were already considered in Sect. 2.3.1, our Lagrangian contains a term that describes the retarded electron-electron interaction. Let $\vec{p}_i$ be the momentum with z-component $p_{i,z}$ and γ_i the γ-factor of the i-th particle for all $n \in \mathbb{N}$ particles, i.e. $i \in \{1, \ldots, n\}$. Thus the Lagrangian can be approximated by (see e.g. [16])

$$L \approx \sum_{i=1}^{n} \left[-\frac{1}{\gamma_i} - q_i \varphi(\vec{r}_i) + q_i \vec{v}_i \cdot \vec{A}(\vec{r}_i) \right] - \sum_{i>j} \left(1 - \frac{p_{i,z} p_{j,z}}{\gamma_i \gamma_j} \right) q_i \varphi_{i,j}. \qquad (3.11)$$

Note that the second sum contains only particles with $i > j$ to avoid counting twice the scalar Liénard-Wiechert potential between two particles, which is denoted by $\varphi_{i,j}$.

Treating the interaction as perturbation, like it is done in [54], the total energy of this system can be described by the Hamiltonian

$$\mathcal{H} = \sum_{i=1}^{n} \left[\gamma_i + q_i \Psi(\vec{r}_i) - p_{i,z} + \sum_{i>j} \left(1 - \frac{p_{i,z} p_{j,z}}{\gamma_i \gamma_j} \right) q_i \varphi_{i,j} \right]. \qquad (3.12)$$

Roughly treated the setting is that there are charged particles, where the i-th particle has rest energy $E_{0,i}$, is subject to an external potential $\Psi_{\text{ext},i}$ and interacts with other particles by means of (retarded) potentials $\varphi_{i,j}$. Thus, the energy of this considered system can be described by the Hamiltonian

$$\mathcal{H} = \sum_{i} \left[E_{0,i} + \Psi_{\text{ext},i} + \sum_{i>j} \varphi_{i,j} \right]. \qquad (3.13)$$

The aim is to find the locations $r_i = (x_i, y_i, z_i)$, where this Hamiltonian is minimal which corresponds to the equilibrium state. Since $E_{0,i}$ is independent of the location and assuming an external force with $\Psi_{\text{ext},i} \propto \|r_i\|^2$ this reduces to find the minimum of

$$\sum_{i} \left[\|r_i\|^2 + \sum_{i>j} \varphi_{i,j} \right]. \qquad (3.14)$$

With the Dirac distribution δ the scalar Liénard-Wiechert potential is essentially

$$\int_{-\infty}^{\infty} \frac{\delta(t - t' - \frac{1}{c} \|r - \tilde{r}(t')\|)}{\|r - \tilde{r}(t')\|} \, dt'. \qquad (3.15)$$

This leads to the question how to handle the discontinuity occurring in the formula. As a first approach the problem is reduced to the z-axis and the discontinuity in z direction modeled by the Heaviside step function θ, i.e. the first goal is to minimize

$$\sum_i \left[\|r_i\|^2 - \sum_{i>j} \frac{\theta(z_i - z_j)}{\|r_i - r_j\|} \right]. \tag{3.16}$$

It would be desirable to develop a step size algorithm to solve such a relativistic problem that works better than conventional methods. In the following, two approaches are discussed. The first approach is to relax the Heaviside step function by the sigmoid function

$$\sigma : \mathbb{R} \to]0, 1[, \ x \mapsto \left(1 + e^{-x}\right)^{-1}. \tag{3.17}$$

The sigmoid function is connected with the hyperbolic tangent by

$$\begin{aligned}
\tfrac{1}{2}(1 + \tanh(x/2)) &= \frac{1}{2}\left(1 + \frac{\sinh(x/2)}{\cosh(x/2)}\right) = \frac{1}{2}\left(1 + \frac{\frac{1}{2}(e^{x/2} - e^{-x/2})}{\frac{1}{2}(e^{x/2} + e^{-x/2})}\right) \\
&= \frac{1}{2}\left(1 + \frac{e^{x/2} + e^{x/2} - (e^{x/2} + e^{-x/2})}{e^{x/2} + e^{-x/2}}\right) = \frac{1}{2}\left(1 + \frac{2e^{x/2}}{e^{x/2} + e^{-x/2}} - 1\right) \\
&= \frac{e^{x/2}}{e^{x/2} + e^{-x/2}} = \frac{1}{1 + e^{-x}} = \sigma(x).
\end{aligned}$$

Thus, the same procedure is possible using $\tanh(x)$ instead of $\sigma(x)$. With $\alpha > 0$, the derivative of $\sigma(\alpha x)$ is

$$\begin{aligned}
\frac{\mathrm{d}}{\mathrm{d}x}\sigma(\alpha x) &= \frac{\mathrm{d}}{\mathrm{d}x}\frac{1}{1 + e^{-\alpha x}} = \frac{\alpha e^{-\alpha x}}{(1 + e^{-\alpha x})^2} = \alpha \cdot \frac{1}{1 + e^{-\alpha x}} \cdot \frac{e^{-\alpha x}}{1 + e^{-\alpha x}} \\
&= \alpha \cdot \frac{1}{1 + e^{-\alpha x}}\left(1 - \frac{1}{1 + e^{-\alpha x}}\right) = \alpha \cdot \sigma(\alpha x)[1 - \sigma(\alpha x)].
\end{aligned}$$

This can be used to determine the slope of $\sigma(\alpha x)$ at $x = 0$, which is given by

$$\frac{\mathrm{d}}{\mathrm{d}x}\sigma(\alpha x)\Big|_{x=0} = \alpha \cdot \tfrac{1}{2}(1 - \tfrac{1}{2}) = \tfrac{\alpha}{4}, \tag{3.18}$$

i.e. the greater α, the steeper the slope in the turning point of $\sigma(\alpha x)$. Since

$$\lim_{\alpha \to \infty} \sigma(\alpha x) = \theta(x), \tag{3.19}$$

the deterioration of the convergence behavior of an optimization algorithm like the gradient descent with Armijo Line Search for increasing α could be examined applied to the objective function

$$\mathcal{H} = \sum_i \left[\|r_i\|^2 - \sum_{i>j} \frac{\sigma(\alpha(z_i - z_j))}{\|r_i - r_j\|} \right]. \qquad (3.20)$$

The second approach is to use a mixed-integer nonlinear programming (MINLP) algorithm (see [55]). For this purpose the problem of minimizing (3.16) needs to be reformulated.

3.3   MINLP-Formulation for Two Particles

For simplicity we consider just two particles, i. e. $n = 2$. Thus, the problem simplifies to the minimization of

$$\|r_1\|^2 + \|r_2\|^2 - \frac{\theta(z_2 - z_1)}{\|r_1 - r_2\|}. \qquad (3.21)$$

Using the definitions

$$\tilde{r} := \begin{pmatrix} r_1 \\ r_2 \end{pmatrix} = \begin{pmatrix} x_1 \\ y_1 \\ z_1 \\ x_2 \\ y_2 \\ z_2 \end{pmatrix}, \quad a := \begin{pmatrix} 0 \\ 0 \\ -1 \\ 0 \\ 0 \\ 1 \end{pmatrix} \quad \text{and} \quad A := \begin{pmatrix} 1 & 0 & 0 & -1 & 0 & 0 \\ 0 & 1 & 0 & 0 & -1 & 0 \\ 0 & 0 & 1 & 0 & 0 & -1 \end{pmatrix} \qquad (3.22)$$

the problem can be written as

$$\min_{\tilde{r}} \ \|\tilde{r}\|^2 - \frac{\theta(a^{\mathrm{T}}\tilde{r})}{\|A\tilde{r}\|}. \qquad (3.23)$$

Assuming the existence of an upper bound U with $|a^{\mathrm{T}}\tilde{r}| \leq U$ this problem can be reformulated equivalently with an additional degree of freedom ξ:

$$\min_{\tilde{r},\xi} \ \left\{ \|\tilde{r}\|^2 - \frac{\xi}{\|A\tilde{r}\|} \ \middle| \ \xi \in \{0, 1\}, \ a^{\mathrm{T}}\tilde{r} \geq -U + \xi U \right\}. \qquad (3.24)$$

Two cases have to be distinguished:

1. If $a^\mathsf{T}\tilde{r} < 0$, i.e. $\theta(a^\mathsf{T}\tilde{r}) = 0$, the restrictions of (3.24) can only be fulfilled by $\xi = 0$.
2. If $a^\mathsf{T}\tilde{r} \geq 0$, i.e. $\theta(a^\mathsf{T}\tilde{r}) = 1$, the restrictions of (3.24) can be fulfilled by both $\xi = 0$ and $\xi = 1$. Nevertheless, since the aim is to minimize the objective function, the optimal choice would be $\xi = 1$.

Therefore, (3.23) and (3.24) have the same optimal value. On the one hand, if $\tilde{r}^*$ denotes the optimal solution of (3.23), $(\bar{\tilde{r}}, \bar{\tilde{\xi}}) = (\tilde{r}^*, \theta(a^\mathsf{T}\tilde{r}^*))$ is the optimal solution of (3.24). On the other hand, if $\tilde{r}^*$, ξ^* is the optimal solution of (3.24), $\tilde{r}^*$ is the optimal solution of (3.23).

Discussion of some Optimization Algorithms

4

Our first approach for minimizing a differentiable function $f : \mathbb{R}^n \to \mathbb{R}$ is using the **Gradient Descent Method (GDM)**. Starting with $x^0 \in \mathbb{R}^n$ the iterates for $k \geq 0$ are given by

$$x^{k+1} = x^k - h_k \nabla f(x^k) \tag{4.1}$$

with positive step sizes h_k. As discussed for example in Chapter 1.2.3 of [56], there are several possibilities to choose the step sizes h_k. For instance, for the minimization of convex functions it can be sufficient to choose a constant step size $h_k \equiv h > 0$. Slightly more advanced it can be assumed to be monotonously decreasing like $h_k = h/\sqrt{k+1}$.

A mostly theoretical approach is to choose the step size by an exact line search, i.e. by

$$h_k = \arg \min_{h \geq 0} f(x^k - h \nabla f(x^k)). \tag{4.2}$$

Since even in one-dimensional cases it can not be guaranteed to find the exact minimum in finite time, it is rarely used in practice. The third approach discussed in [56] is the Armijo rule, which is analyzed in more detail in the next chapter.

M. Hagedorn, *Minimization Problems for the Witness Beam in Relativistic Plasma Cavities*, BestMasters, https://doi.org/10.1007/978-3-658-46226-0_4

4.1   Armijo Line Search

We fix parameters α and β with

$$0 < \alpha < \beta < 1 \tag{4.3}$$

and find $x^{k+1} = x^k - h\nabla f(x^k)$ with $h > 0$ such that

$$\alpha\left\langle \nabla f(x^k),\, x^k - x^{k+1}\right\rangle \leq f(x^k) - f(x^{k+1}), \tag{4.4}$$

$$\beta\left\langle \nabla f(x^k),\, x^k - x^{k+1}\right\rangle \geq f(x^k) - f(x^{k+1}). \tag{4.5}$$

Let $x^k \in \mathbb{R}^n$ be fixed with $\nabla f(x^k) \neq 0$. We define

$$\phi:\ \mathbb{R}_+ \to \mathbb{R},\ h \mapsto \phi(h) := f(x^k - h\nabla f(x^k)), \tag{4.6}$$

what we already know as objective function for the exact line search. The first condition of the Armijo rule (4.4) can be written as

$$\alpha\left\langle \nabla f(x^k),\, x^k - x^{k+1}\right\rangle \leq f(x^k) - f(x^{k+1})$$

$$\Leftrightarrow\ \alpha\left\langle \nabla f(x^k),\, x^k - (x^k - h\nabla f(x^k))\right\rangle \leq f(x^k) - f(x^{k+1})$$

$$\Leftrightarrow\ \alpha h\left\langle \nabla f(x^k),\, \nabla f(x^k))\right\rangle \leq f(x^k) - f(x^{k+1})$$

$$\Leftrightarrow\ f(x^{k+1}) \leq \underbrace{f(x^k) - \alpha h\|\nabla f(x^k)\|^2}_{=:\phi_\alpha(h)}$$

$$\Leftrightarrow\ f(x^{k+1}) \leq \phi_\alpha(h)$$

and the second condition (4.5) analogously

$$\beta\left\langle \nabla f(x^k),\, x^k - x^{k+1}\right\rangle \geq f(x^k) - f(x^{k+1})$$

$$\Leftrightarrow\ f(x^{k+1}) \geq \underbrace{f(x^k) - \beta h\|\nabla f(x^k)\|^2}_{=:\phi_\beta(h)}$$

$$\Leftrightarrow\ f(x^{k+1}) \geq \phi_\beta(h).$$

Note that

$$\phi(0) = f(x^k) = \phi_\alpha(0) = \phi_\beta(0) \tag{4.7}$$

and since the derivatives are given by

$$\phi'(h)\Big|_{h=0} = f'(x^k - h\nabla f(x^k)) \cdot (-\nabla f(x^k))\Big|_{h=0} = -\|\nabla f(x^k)\|^2 < 0,$$

$$\phi_\alpha'(h)\Big|_{h=0} = \phi_\alpha'(h) = -\alpha\|\nabla f(x^k)\|^2,$$

$$\phi_\beta'(h)\Big|_{h=0} = \phi_\beta'(h) = -\beta\|\nabla f(x^k)\|^2,$$

we have

$$\phi'(0) < \phi_\beta'(0) < \phi_\alpha'(0) < 0, \tag{4.8}$$

because of the condition $0 < \alpha < \beta < 1.$

To illustrate the geometric interpretation of this Armijo rule we consider the following example. The aim is to minimize the two-dimensional function

$$f : \mathbb{R}^2 \to \mathbb{R}, \ (x, y) \mapsto f(x, y) := \cos(x)\cos(y). \tag{4.9}$$

Starting with $(x^0, y^0)^\mathsf{T} := (-2.5, -3.5)^\mathsf{T}$ with $f(x^0, y^0) \approx 0.75$ and descent direction

$$-\nabla f(x^0, y^0) \approx (0.56, -0.28)^\mathsf{T} \neq 0 \tag{4.10}$$

the setting is shown in Fig. 4.1.

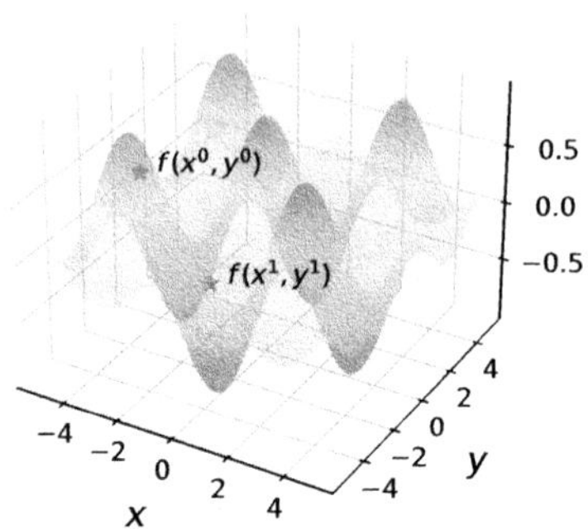

Figure 4.1 Three-dimensional plot of the function $f(x, y) = \cos(x)\cos(y)$

In our example, $\phi(h)$, $\phi_\alpha(h)$ and $\phi_\beta(h)$ are given by

$$\phi(h) = f((x^0, y^0)^\mathrm{T} - h\nabla f(x^0, y^0)), \tag{4.11}$$

$$\phi_\alpha(h) = f(x^0, y^0) - \alpha h \|\nabla f(x^0, y^0)\|^2, \tag{4.12}$$

$$\phi_\beta(h) = f(x^0, y^0) - \beta h \|\nabla f(x^0, y^0)\|^2 \tag{4.13}$$

From Fig. 4.2 we can see that ϕ is intersected by ϕ_β and ϕ_α at $h \approx 6.1$ and $h \approx 6.6$ respectively. Thus, for example the choice $h_0 = 6.3$ meets the conditions (4.4) and (4.5) of the Armijo rule. This choice leads to the new coordinates $(x^1, y^1)^\mathrm{T} \approx (1.0, -5.3)^\mathrm{T}$ with $f(x^1, y^1) \approx 0.27$ as marked in Fig. 4.1.

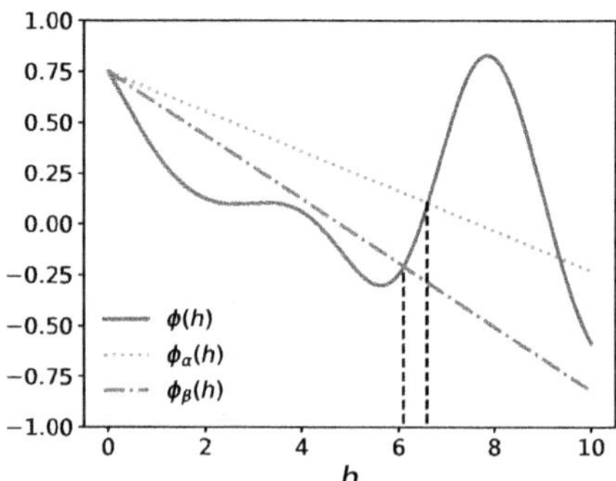

Figure 4.2 Plot of the functions $\phi(h)$, $\phi_\alpha(h)$ and $\phi_\beta(h)$ with the choice $\alpha = 0.25$ and $\beta = 0.4$

4.2   Disadvantage of using the Gradient Descent Method

Using the notation (3.22), for $n = 2$ the Hamiltonian (3.20) reduces to

$$\sum_i \left[\|r_i\|^2 - \sum_{i>j} \frac{\sigma(\alpha(z_i - z_j))}{\|r_i - r_j\|} \right] = \|r_1\|^2 + \|r_2\|^2 - \frac{\sigma(\alpha(z_2 - z_1))}{\|r_1 - r_2\|} = \|\tilde{r}\|^2 - \frac{\sigma(\alpha(a^{\mathrm{T}}\tilde{r}))}{\|A\tilde{r}\|}.$$

$$(4.14)$$

In case of $n = 3$, the definitions

$$\tilde{r} := (r_1, r_2, r_3)^{\mathrm{T}} = (x_1, y_1, z_1, x_2, y_2, z_2, x_3, y_3, z_3)^{\mathrm{T}}, \quad a_1 := (0, 0, -1, 0, 0, 1, 0, 0, 0)^{\mathrm{T}},$$

$$a_2 := (0, 0, -1, 0, 0, 0, 0, 0, 1)^{\mathrm{T}}, \quad a_3 := (0, 0, 0, 0, 0, 0, -1, 0, 0, 1)^{\mathrm{T}},$$

$$A1 := (I_3, -I_3, 0_3), \quad A2 := (I_3, 0_3, -I_3), \quad \text{and} \quad A3 := (0_3, I_3, -I_3)$$

with $n \times n$ identity matrix I_n and zero matrix 0_n help to simplify the Hamiltonian:

$$\mathcal{H} = \sum_i \left[\|r_i\|^2 - \sum_{i>j} \frac{\sigma(\alpha(z_i - z_j))}{\|r_i - r_j\|} \right]$$

$$= \|r_1\|^2 + \|r_2\|^2 + \|r_3\|^2 - \frac{\sigma(\alpha(z_2 - z_1))}{\|r_1 - r_2\|} - \frac{\sigma(\alpha(z_3 - z_1))}{\|r_1 - r_3\|} - \frac{\sigma(\alpha(z_3 - z_2))}{\|r_2 - r_3\|}$$

$$= \|\tilde{r}\|^2 - \frac{\sigma(\alpha(a_1^{\mathrm{T}}\tilde{r}))}{\|A_1\tilde{r}\|} - \frac{\sigma(\alpha(a_2^{\mathrm{T}}\tilde{r}))}{\|A_2\tilde{r}\|} - \frac{\sigma(\alpha(a_3^{\mathrm{T}}\tilde{r}))}{\|A_3\tilde{r}\|} = \|\tilde{r}\|^2 - \sum_{i=1}^{3} \frac{\sigma(\alpha(a_i^{\mathrm{T}}\tilde{r}))}{\|A_i\tilde{r}\|}.$$

While $\|\tilde{r}\|^2$ is minimal for $\tilde{r} = 0$, the other terms tend to infinity for $\tilde{r} \to 0$. Of course, a realistic particle number n is significantly larger than 2 or 3, but the increase of the number of interaction terms by increasing the number of particles can be imagined. For an arbitrary number of particles $n \in \mathbb{N}$ the number of interaction terms is given by

$$\sum_{k=1}^{n-1} k = \frac{(n-1) \cdot n}{2}.$$

$$(4.15)$$

As indicated in the title of this chapter, the aim here is to investigate whether the Gradient Descent Method (also called Steepest Descent Method) is suitable for minimizing our Hamiltonian. The Gradient Descent Method is an iterative method for minimizing a differentiable function $f : \mathbb{R}^n \to \mathbb{R}$, which provides a sequence of iterates $(x^k)_{k \in \mathbb{N}_0}$ for a starting value x^0 and a step size $h_k > 0$ following the update rule

$$x^{k+1} := x^k - h_k \nabla f(x^k). \tag{4.16}$$

In the case $n = 3$, the gradient of our Hamiltonian is given by

$$\nabla_{\tilde{r}} \mathcal{H} = 2\tilde{r} - \sum_{i=1}^{3} \frac{\alpha \|A_i \tilde{r}\| \sigma(\alpha(a_i^T \tilde{r}))[1 - \sigma(\alpha(a_i^T \tilde{r}))]a_i - \sigma(\alpha(a_i^T \tilde{r}))\|A_i \tilde{r}\|^{-1} A_i \tilde{r}}{\|A_i \tilde{r}\|^2}.$$

$$\tag{4.17}$$

Let us now take a closer look at terms of the type $\sigma(\alpha x)$, as they occur in the interaction terms.

Proposition 4.1 *A doubling of $\alpha > 0$ leads roughly to a halving of the support of the derivative of the sigmoid function*

$$\sigma_\alpha : \mathbb{R} \to \,]0, 1[\,, \quad \sigma_\alpha(x) := \sigma(\alpha x). \tag{4.18}$$

Proof. Since the first four derivatives are given by

$$\sigma_\alpha'(x) = \alpha\sigma_\alpha(x)[1 - \sigma_\alpha(x)], \quad \sigma_\alpha''(x) = \alpha\sigma_\alpha'(x)[1 - 2\sigma_\alpha(x)],$$

$$\sigma_\alpha'''(x) = \alpha^2 \sigma_\alpha'(x)[1 - 2\sigma_\alpha(x)]^2 - 2\alpha(\sigma_\alpha'(x))^2, \quad \text{and}$$

$$\sigma_\alpha^{(4)}(x) = \alpha^2 \sigma_\alpha''(x)[1 - 2\sigma_\alpha(x)]^2 - 4\alpha^2(\sigma_\alpha'(x))^2[1 - 2\sigma_\alpha(x)] - 4\alpha\sigma_\alpha'(x)\sigma_\alpha''(x),$$

and have the intercepts

$$\sigma_\alpha(0) = \tfrac{1}{2}, \quad \sigma_\alpha'(0) = \tfrac{\alpha}{4}, \quad \sigma_\alpha''(0) = 0, \quad \sigma_\alpha'''(0) = \tfrac{-\alpha^3}{8}, \quad \text{and } \sigma_\alpha^{(4)}(0) = 0,$$

the Taylor expansion of $\sigma_\alpha(x)$ at $x = 0$ to the fourth order is

$$\sigma_\alpha(x) = \sum_{k=0}^{\infty} \frac{\sigma_\alpha^{(k)}(0)}{k!} x^k = \tfrac{1}{2} + \tfrac{\alpha}{4}x + 0x^2 - \tfrac{1}{3!}\tfrac{\alpha^3}{8}x^3 + 0x^4 + \mathcal{O}(x^5) = \tfrac{1}{2} + \tfrac{\alpha}{4}x - \tfrac{\alpha^3}{48}x^3 + \mathcal{O}(x^5).$$

We approximate the support of σ_α by the distance between the extrema of the Taylor expansion $f : \mathbb{R} \to \mathbb{R}$, $f(x) := \tfrac{1}{2} + \tfrac{\alpha}{4}x - \tfrac{\alpha^3}{48}x^3$.

Necessary condition: $f'(x) = \tfrac{\alpha}{4} - \tfrac{\alpha^3}{16}x^2 \overset{!}{=} 0 \Rightarrow \tfrac{16\alpha}{4\alpha^3} = \tfrac{4}{\alpha^2} \overset{!}{=} x^2$, i.e. $x \in \{-\tfrac{2}{\alpha}, \tfrac{2}{\alpha}\}$.

Sufficient condition: $f''(\pm\tfrac{2}{\alpha}) = -\tfrac{\alpha^3}{8} \cdot (\pm\tfrac{2}{\alpha}) = \mp\tfrac{\alpha^2}{4} \lessgtr 0$.

Outside the interval $[-\frac{2}{\alpha}, \frac{2}{\alpha}]$, the slope is close to zero and is therefore neglected in our estimation. Thus, the distance between the minimum at $x = -\frac{2}{\alpha}$ and the maximum at $x = \frac{2}{\alpha}$ is

$$\left| \frac{2}{\alpha} - \left(-\frac{2}{\alpha}\right) \right| = \frac{4}{\alpha},$$

from which the proposition follows directly. $\qquad\square$

The above result is visualized in Fig. 4.3:

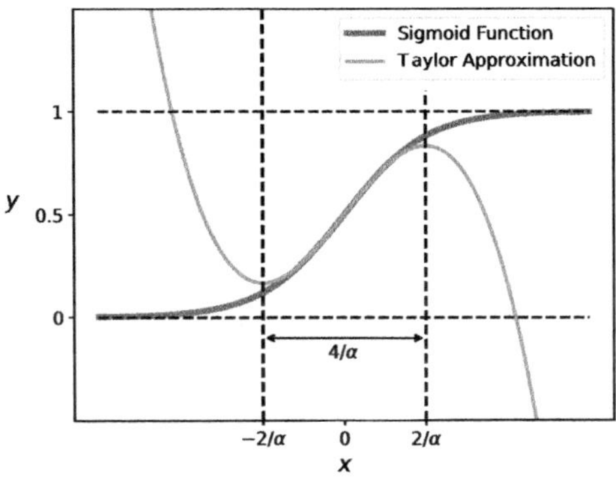

Figure 4.3 Sigmoid function $\sigma_\alpha(x)$ and its Taylor expansion to the fourth order

Since the sigmoid function was introduced as an approximation to the Heaviside step function, an appropriate choice for α is relatively large and thus the supports of the interaction terms in the gradient of the Hamiltonian (4.17) are correspondingly small. Therefore, the interactions are hardly taken into account by the Gradient Descent Method (4.16).

An alternative approach would be to approximate the Heaviside step function $\theta(x)$ by a piecewise defined, continuously differentiable function f that is described for a fixed $\varepsilon > 0$ on $[-\varepsilon, \ \varepsilon]$ by a cubic polynomial $p(x) = ax^3 + bx^2 + cx + d$. The real parameters $a, \ b, \ c$ and d have to be chosen in such a way, so that

$$f : \mathbb{R} \to [0, 1], \ x \mapsto f(x) := \begin{cases} 0 & , x < -\varepsilon \\ p(x) & , x \in [-\varepsilon, \ \varepsilon] \\ 1 & , x > \varepsilon \end{cases} \qquad (4.19)$$

is continuously differentiable, i. e.

$$p(-\varepsilon) \stackrel{!}{=} 0, \;\; p(\varepsilon) \stackrel{!}{=} 1 \;\; \text{and} \;\; p'(-\varepsilon) \stackrel{!}{=} p'(\varepsilon) \stackrel{!}{=} 0, \tag{4.20}$$

where the derivative of the polynomial is given by $p'(x) = 3ax^2 + 2bx + c$. The conditions in (4.20) lead to the following system of linear equations:

$$\begin{cases} -a\varepsilon^3 + b\varepsilon^2 - c\varepsilon + d = 0 \\ a\varepsilon^3 + b\varepsilon^2 + c\varepsilon + d = 1 \\ 3a\varepsilon^2 - 2b\varepsilon + c = 0 \\ 3a\varepsilon^2 + 2b\varepsilon + c = 0 \end{cases} \Leftrightarrow \begin{pmatrix} -\varepsilon^3 & \varepsilon^2 & -\varepsilon & 1 \\ \varepsilon^3 & \varepsilon^2 & \varepsilon & 1 \\ 3\varepsilon^2 & -2\varepsilon & 1 & 0 \\ 3\varepsilon^2 & 2\varepsilon & 1 & 0 \end{pmatrix} \begin{pmatrix} a \\ b \\ c \\ d \end{pmatrix} = \begin{pmatrix} 0 \\ 1 \\ 0 \\ 0 \end{pmatrix}.$$

It is solved by

$$\begin{pmatrix} a \\ b \\ c \\ d \end{pmatrix} = \begin{pmatrix} -\varepsilon^3 & \varepsilon^2 & -\varepsilon & 1 \\ \varepsilon^3 & \varepsilon^2 & \varepsilon & 1 \\ 3\varepsilon^2 & -2\varepsilon & 1 & 0 \\ 3\varepsilon^2 & 2\varepsilon & 1 & 0 \end{pmatrix}^{-1} \begin{pmatrix} 0 \\ 1 \\ 0 \\ 0 \end{pmatrix} = \begin{pmatrix} \frac{1}{4\varepsilon^3} & \frac{-1}{4\varepsilon^3} & \frac{1}{4\varepsilon^2} & \frac{1}{4\varepsilon^2} \\ 0 & 0 & \frac{-1}{4\varepsilon} & \frac{1}{4\varepsilon} \\ \frac{-3}{4\varepsilon} & \frac{3}{4\varepsilon} & \frac{-1}{4} & \frac{-1}{4} \\ \frac{1}{2} & \frac{1}{2} & \frac{\varepsilon}{4} & \frac{-\varepsilon}{4} \end{pmatrix} \begin{pmatrix} 0 \\ 1 \\ 0 \\ 0 \end{pmatrix} = \begin{pmatrix} \frac{-1}{4\varepsilon^3} \\ 0 \\ \frac{3}{4\varepsilon} \\ \frac{1}{2} \end{pmatrix},$$

which leads to the polynomial $p(x) = \frac{-1}{4\varepsilon^3}x^3 + \frac{3}{4\varepsilon}x + \frac{1}{2}$ for a given $\varepsilon > 0$. The comparison between Fig. 4.3 and Fig. 4.4 shows that this second approach is a better approximation for the Heaviside step function. Nevertheless, the result that a doubling of the slope at the zero point corresponds to a halving of the support of the derivative can be confirmed.

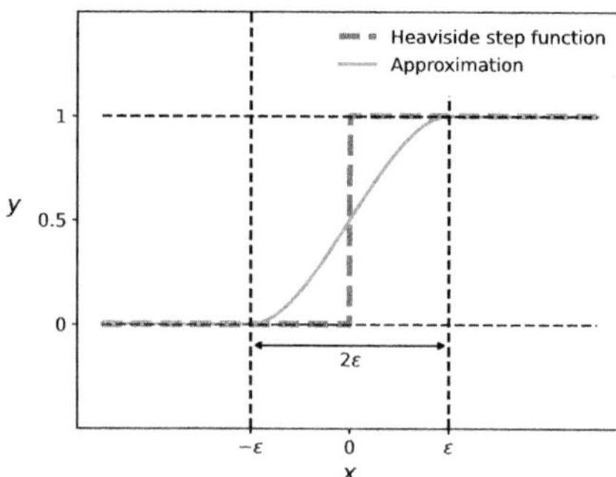

Figure 4.4 Heaviside step function and its approximation by a piecewise defined, continuously differentiable function f

4.3   Heavy Ball Method and Nesterov's Accelerated Gradient Descent

To accelerate the convergence, it can be added a kind of short term memory to the steepest descent method, i.e. in addition to the current iterate x^k and gradient $\nabla f(x^k)$ one can also use the previous iterate x^{k-1} or gradient $\nabla f(x^{k-1})$ to calculate the next iterate x^{k+1}.

The first modification discussed here is the **Heavy Ball Method (HBM)**, which was proposed by Polyak in [57]. With the initialization $x^{-1} := x^0 \in \mathbb{R}^n$, a step size $\alpha > 0$ and a parameter $\beta \in \,]0, 1)$ the update rule is

$$x^{k+1} = x^k - \alpha \nabla f(x^k) + \beta(x^k - x^{k-1}). \tag{4.21}$$

Defining $m^k := \frac{\beta}{\alpha}(x^k - x^{k-1}) - \nabla f(x^k)$ with $m^0 := -\nabla f(x^0)$ leads to

$$x^{k+1} = x^k - \alpha \nabla f(x^k) + \beta(x^k - x^{k-1}) = x^k + \alpha\left(\frac{\beta}{\alpha}(x^k - x^{k-1}) - \nabla f(x^k)\right) = x^k + \alpha m^k$$

$$\text{with } m^{k+1} = \frac{\beta}{\alpha}(x^{k+1} - x^k) - \nabla f(x^{k+1}) = \frac{\beta}{\alpha}\alpha m^k - \nabla f(x^{k+1}) = \beta m^k - \nabla f(x^{k+1}),$$

also known as momentum method.

The second modification is **Nesterov's Accelerated Gradient Method (NAG)** [56]. With the initialization $y^0 := x^0 \in \mathbb{R}^n$ and suitable parameters $\alpha, \beta > 0$ the sequence is defined by

$$x^{k+1} = y^k - \alpha \nabla f(y^k) \quad \text{with} \quad y^{k+1} = x^{k+1} + \beta(x^{k+1} - x^k). \tag{4.22}$$

With $x^{-1} := x^0$ this is equivalent to

$$x^{k+1} = y^k - \alpha \nabla f(y^k) = x^k + \beta(x^k - x^{k-1}) - \alpha \nabla f(x^k + \beta(x^k - x^{k-1})).$$

Note, that in (HBM) as well as in (NAG) the choice $\beta = 0$ would simply result in the steepest descent method with constant step size α.

4.4   Relativistic Gradient Descent

In [58], a generalization of the two discussed modifications of the steepest descent method is proposed, which could accelerate and stabilize the optimization process without additional computational costs. This so-called **Relativistic Gradient Descent (RGD)** for minimizing a differentiable function $f : \mathbb{R}^n \to \mathbb{R}$ is proposed as follows:

Algorithm 4.2
   Initialization: $x^0, v^0 \in \mathbb{R}^n$, $\alpha > 0$, $\delta > 0$, $\beta \in\,]0, 1[\,$, $\eta \in [0, 1]$.
   for $k \in \{0, 1, \dots\}$ ***do***

$$x^{k+1/2} \leftarrow x^k + \sqrt{\beta}v^k / \sqrt{\beta\delta\|v^k\|^2 + 1}$$

$$v^{k+1/2} \leftarrow \sqrt{\beta}v^k - \alpha\nabla f(x^{k+1/2})$$

$$x^{k+1} \leftarrow \eta x^{k+1/2} + (1 - \eta)x^k + v^{k+1/2} / \sqrt{\delta\|v^{k+1/2}\|^2 + 1}$$

$$v^{k+1} \leftarrow \sqrt{\beta}v^{k+1/2}$$

   end for

With $\delta = 0$ and $\eta = 0$ Algorithm 4.2 simplifies to

for $k \in \{0, 1, \dots\}$ **do**

$$x^{k+1/2} \leftarrow x^k + \sqrt{\beta}v^k$$

$$v^{k+1/2} \leftarrow \sqrt{\beta}v^k - \alpha\nabla f(x^{k+1/2}) = \sqrt{\beta}v^k - \alpha\nabla f(x^k + \sqrt{\beta}v^k)$$

$$x^{k+1} \leftarrow x^k + v^{k+1/2} = x^k + \sqrt{\beta}v^k - \alpha\nabla f(x^k + \sqrt{\beta}v^k)$$

$$v^{k+1} \leftarrow \sqrt{\beta}v^{k+1/2}$$

end for

Thus, (RGD) recovers (NAG).

And for $\delta = 0$ and $\eta = 1$ we get

for $k \in \{0, 1, \dots\}$ **do**

$$x^{k+1/2} \leftarrow x^k + \sqrt{\beta}v^k$$

$$v^{k+1/2} \leftarrow \sqrt{\beta}v^k - \alpha\nabla f(x^{k+1/2}) = \sqrt{\beta}v^k - \alpha\nabla f(x^k + \sqrt{\beta}v^k)$$

$$x^{k+1} \leftarrow x^{k+1/2} + v^{k+1/2} = x^{k+1/2} + \sqrt{\beta}v^k - \alpha\nabla f(x^{k+1/2})$$

$$v^{k+1} \leftarrow \sqrt{\beta}v^{k+1/2}$$

end for

which means, that (RGD) becomes a second order accurate version of (HBM).

Defining $y^k := \eta x^{k+1/2} + (1 - \eta)x^k$ leads to

$$x^{k+1} := \eta x^{k+1/2} + (1 - \eta)x^k + v^{k+1/2}/\sqrt{\delta\|v^{k+1/2}\|^2 + 1}$$

$$= y^k + v^{k+1/2}/\sqrt{\delta\|v^{k+1/2}\|^2 + 1}$$

and therefore

$$\|x^{k+1} - y^k\|^2 = \frac{\|v^{k+1/2}\|^2}{\delta\|v^{k+1/2}\|^2 + 1} = \begin{cases} \frac{1}{\delta+1/\|v^{k+1/2}\|^2} \leq \frac{1}{\delta} & \text{if } v^{k+1/2} \neq 0 \\ 1 & \text{if } v^{k+1/2} = 0. \end{cases}$$

Thus, (RGD) is globally bounded independently of the norm of the gradient. In contrast, for (HBM) and (NAG) we have $\delta = 0$, i.e. $\|x^{k+1} - y^k\|^2 \leq \infty$.

Note that there seems to be a typo in [58]. On page 2 there is written $\|x_{k+1} - y_k\| \leq 1/\delta$. Either a square or a square root needs to be added as it is derived above.

4.5   Algorithm for Hyperparameter Optimization

To find the best choice of parameters for the algorithms discussed in this chapter, we use the function `min_f` suggested by [59]. The provided MATLAB-code was translated to Python for this thesis and can be found on [60].

The function `min_f` can be used for the minimization of a smooth, at least once differentiable function $f : \mathbb{R}^n \to \mathbb{R}$. The derivative of f does not have to be known and f is allowed to be nonlinear and non-convex. It is possible to limit the search range by specifying upper and lower bounds.

The algorithm is particularly suitable when the function evaluations are cheap and when the rounding errors are small. Furthermore, the dimension should be at most 300, since the number of needed iterations grows moderately with n. The aim of the algorithm is to find a local minimum. No additional effort is made to find the global minimum. It is therefore advisable to choose sufficiently good initial estimates.

`min_f` uses a Quasi-Newton trust region approach (see e. g. [61]) with the Powell symmetric Broyden's method followed by a curvature correction. The trust region radius is adjusted with a line search, which is based on a spline function. This spline function minimizes a weighted least squares sum, where the weights penalize large changes of the third derivative of the spline function between short intervals more than large changes between longer intervals.

Since $\texttt{min\_f}$ does not use any information of the derivatives of f, the gradients need to be approximated. For this purpose, central finite differences are used, which are very expensive to calculate. Further applied concepts are a randomized basis approach and the active set strategy.

In summary, $\texttt{min\_f}$ is a method that is not particularly efficient and requires sufficiently good initial values in order to find the global minimum. Nevertheless, it offers the great advantage of being able to minimize arbitrary functions, even if their gradients are not implemented. This is why we decide using $\texttt{min\_f}$ to determine the free parameters of the methods presented in this chapter.

Numerical Simulations 5

For the implementation the Python version 3.9.18, the R version 4.3.1, the MATLAB version R2020b and a laptop with an Intel i7-10510U processor with eight cores and 16GB of RAM were used. The minimization of the Hamiltonian using a simplified version of the Armijo Line Search is a translation of the MATLAB code of [16]. In this chapter, this code is extended so that the modifications of the steepest descent method discussed in Chap. 4 can be used instead of the Armijo line search. After a hyperparameter optimization, which is intended to ensure that all algorithms under consideration perform as well as possible, the algorithms are compared with each other.

5.1 Setting

Initially, we consider a spherical region with radius $R = 50\,\mathrm{nm}$ inside the bubble. We randomly distribute $N = 500$ electrons on a plane perpendicular to the direction of propagation z, using a uniform distribution on this circular area. We choose a co-moving reference frame by defining $\xi := z - V_0 t$, where V_0 is the velocity of the bubble. The wakefield potential Ψ of the i-th particle in our Hamiltonian (3.12) can therefore be expressed by $\Psi(\vec{r}_i) = (x_i^2 + y_i^2 + \xi_i^2)/8$. This setting can look like it is shown in Fig. 5.1:

© The Author(s), under exclusive license to Springer Fachmedien Wiesbaden GmbH, part of Springer Nature 2024
M. Hagedorn, *Minimization Problems for the Witness Beam in Relativistic Plasma Cavities*, BestMasters, https://doi.org/10.1007/978-3-658-46226-0_5

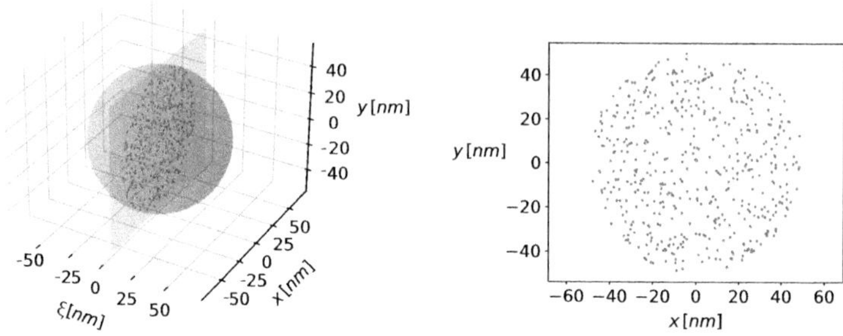

(a) Random distributed particles within a circular area in the bubble.

(b) Two-dimensional projection of Figure 15a.

Figure 5.1 Initial distribution of the particles under consideration

We assume a particle momentum of $p = 250\,\text{MeV}/c$ and a plasma wavelength of $\lambda = 0.01\,\text{cm}$. Minimizing the Hamiltonian using the steepest descent method with Armijo Line Search with tolerance 10^{-8}, initial step size 0.1 and reduction factor 0.9 leads to the regular arrangement of the electrons, which can be seen in Fig. 5.2.

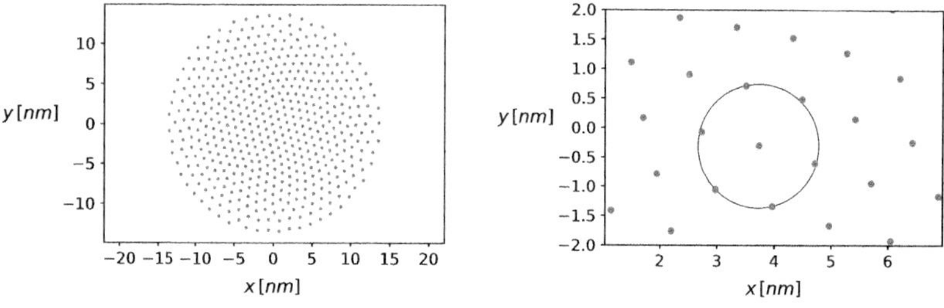

(a) Overview of the electron configuration after the minimization process.

(b) Enlargement of a section of Figure 16a to reveal the hexagonal structure.

Figure 5.2 Particle arrangement after the minimization of the Hamiltonian

The electrons are arranged as Wigner crystals. In [15] it is derived in detail how the interaction between free electrons in an electron gas leads to this structure.

5.2   Parameter Selection

The aim of this subsection is, for example, to determine how α and β have to be selected for the Heavy Ball Method so that the gradient of the Hamiltonian is as small as possible after as few iterations as possible. For this purpose, we use the function `min_f`, which was already discussed in Sect. 4.5.

The questions we would like to answer are: Which function should be minimized? How large are the fluctuations of the optimal parameters depending on the random start configuration? Are there any dependencies on the momentum or on the number of particles?

We start with the momentum $p = 125\,\text{MeV/c}$, the wavelength $\lambda = 0.01\,\text{cm}$ and $N = 100$ electrons and are interested in determining the parameters α and β such that (HBM) leads to a final iterate with a small value for the norm of the gradient of the Hamiltonian. A few attempts by hand motivate a grid search for $\alpha \in \{0.0001,\ 0.0003,\ 0.001,\ 0.003,\ 0.01,\ 0.03,\ 0.1,\ 0.3,\ 1\}$ and $\beta \in \{0,\ 0.1,\ 0.2,\ 0.3,\ 0.4,\ 0.5,\ 0.6,\ 0.7,\ 0.8,\ 0.9,\ 1\}$. The results of this grid search are clearly presented in Fig. 5.3. Where the norm of the final gradient was "not a number" the corresponding field is left blank. As the initial configuration of the electrons is configured randomly to a certain extent, there are fluctuations— particularly in the values for β. For stabilization, ten independent grid searches are carried out and the results are averaged. This is what is actually shown in Fig. 5.3.

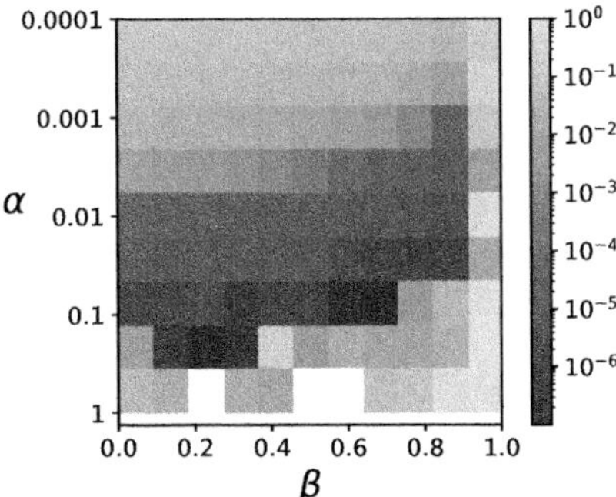

Figure 5.3 Norm of the final gradient obtained by (HBM) depending on the parameters α and β

Unfortunately, Fig. 5.3 displays that (HBM) achieves particularly good results for those parameters that are close to the range in which no result can be determined at all. Careful consideration of these advantages and disadvantages leads to the preliminary recommendation

$$\alpha = 0.3 \ \text{ and } \ \beta = 0.2. \tag{5.1}$$

These recommendations serve as initial values using `min_f`. Again, ten independent runs are performed and the results are presented as boxplots (see Fig. 5.4).

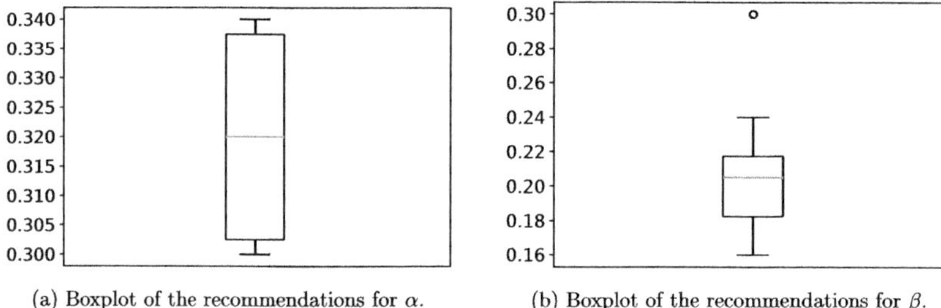

(a) Boxplot of the recommendations for α.   (b) Boxplot of the recommendations for β.

Figure 5.4 Parameters for (HBM) suggested by `min_f` for ten independent runs

Taking into account the boxplots in Fig. 5.4, we can minimally adjust the step size to $\alpha = 0.32$.

Let us now turn our attention to the question of whether a different choice of momentum or wavelength would also result in a different choice of parameters α and β. For this purpose, we repeat the above grid search, varying the settings for the momentum $p \in \{5, 25, 125, 625\}$MeV/c, the wavelength $\lambda \in \{0.001, 0.003, 0.01, 0.03, 0.1\}$cm and the number of electrons

$$N \in \{25, 50, 100, 200, 400, 800\} \tag{5.2}$$

individually. Although the values of the norm of the gradient change, the picture like in Fig. 5.3 remains the same for different values of p, λ, the initial radius or the number of iterations. Nevertheless, different numbers of electrons lead to a vertical shift and therefore larger numbers of particles require smaller step sizes. The numerically obtained scaling law for the step size α is roughly $\alpha \propto \frac{1}{\sqrt{N}}$. Thus, we finally recommend for the Heavy Ball Method (HBM)

$$\alpha = \frac{3}{\sqrt{N}} \quad \text{and} \quad \beta = 0.2, \tag{5.3}$$

which leads to (5.1) for $N = 100$.

The next step is to find appropriate parameters for (NAG). The same procedure leads to

$$\alpha = \frac{1}{\sqrt{N}} \quad \text{and} \quad \beta = 0.8, \tag{5.4}$$

in this case, i.e. to $\alpha = 0.1$ and $\beta = 0.8$ for $N = 100$. Again, this recommendation is independent of the choice of p, λ, the initial radius and the number of iterations.

Four parameters need to be determined so that (RGD) can perform at its best, namely α, β, δ and η. Although the grid search for four parameters is not as easy to visualize as for two parameters, the underlying principle is identical. Based on the parameter selections in the examples provided as supplementary material to [58], we start with a relatively rough grid search for

$$\begin{aligned}
\alpha &\in \{10^{-5},\ 10^{-4},\ 10^{-3},\ 10^{-2},\ 10^{-1}\}, \\
\beta &\in \{0,\ 0.1,\ 0.2,\ 0.3,\ 0.4,\ 0.5,\ 0.6,\ 0.7,\ 0.8,\ 0.9,\ 1\}, \\
\delta &\in \{0.1,\ 0.3,\ 1,\ 3,\ 10,\ 30,\ 100\}, \\
\eta &\in \{0,\ 0.25,\ 0.5,\ 0.75,\ 1\}.
\end{aligned} \tag{5.5}$$

After refining the search with

$$\alpha \in \{0.01,\ 0.02,\ 0.04,\ 0.08,\ 0.16,\ 0.32,\ 0.64\}, \tag{5.6}$$

it was found that the parameter selection is independent of the momentum and the plasma wave length, neglecting statistical fluctuations in the starting configuration. As with the previous algorithms, we observe that smaller step sizes α become necessary as the number of particles increases. In this context,

$$\beta = 0.8,\ \delta = 1,\ \eta = 0.5 \tag{5.7}$$

is identified as a good choice for the remaining parameters, although it should be noted that deviating values also lead to similarly good results.

Now we intend to examine the relationship between α and N in more detail. For

$$N \in \{10, \ 20, \ 40, \ 80, \ 160, \ 320, \ 640, \ 1280\} \tag{5.8}$$

the rough order of magnitude of α is first determined by a simple line search in dependence of N. This is then used as an initial value to determine with `min_f` a more precise value for α by averaging about the results for three independent starting configurations. Fitting a linear regression model leads to the final recommendation to select the parameters approximately as follows:

$$\alpha = \frac{0.9}{\sqrt{N}}, \ \beta = 0.8, \ \delta = 10^{-9}, \ \eta = 0.3. \tag{5.9}$$

5.3   Comparison of Performance

The results that can be achieved with the different algorithms will be compared in this chapter. We fix $p = 125\,\text{MeV/c}$, $\lambda = 0.01\,\text{cm}$ and $N = 100$ and use the parameters which we have determined in Sect. 5.2. For the analysis, we calculate the norm of the gradient of the Hamiltonian in each iteration. While the norm of the gradient is used as a termination criterion in the Gradient Descent Method, it is usually not required in the other methods.

For stability reasons, ten random start configurations are considered independently of each other. All four methods are used for each start configuration. For each method, the arithmetic mean of the twenty results is calculated in each iteration.

In Fig. 5.5 the logarithms of these arithmetic means to the base ten are plotted against the number of iterations. The ends of the added vertical bars are located at the arithmetic mean plus or minus the standard deviation of the logarithmized values.

Figure 5.5 shows that (HBM) is the best choice if the number of iterations is relatively small, since after just a few iterations we get significantly smaller values for the norm of the gradient than with the other algorithms. Shortly before (GDM) stagnates, the results obtained using the other algorithms are only marginally better, but in contrast to (GDM), they continue to improve with increasing number of iterations.

On average, the lowest values are obtained using (RGD), but it should be taken into account that the standard deviation is very large. For individual random start configurations, it is not possible to predict which of the three modifications of (GDM) lead to the lowest values of the norm of the gradient.

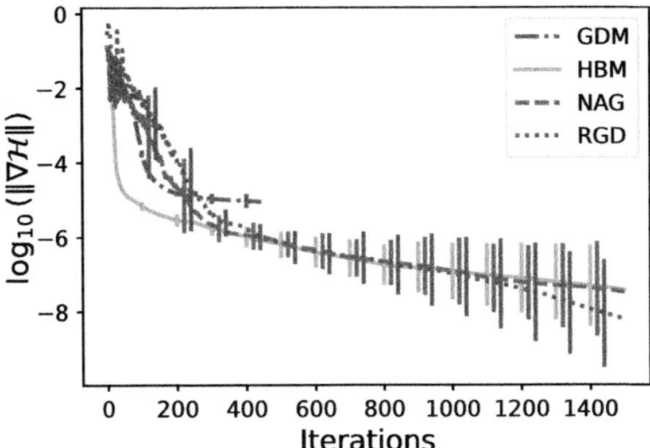

Figure 5.5 Comparison of Gradient Descent Method (GDM), Heavy Ball Method (HBM), Nesterov's Accelerated Gradient Method (NAG) and Relativistic Gradient Descent Method (RGD). In each case the arithmetic mean of ten independent runs was used

Finally, it should be remembered that the iterations of the different methods require different amounts of computing power. For example, (GDM) requires the calculation of the norm of the gradient in each iteration and previous iterations must be saved for the modifications of (GDM). The results from Fig. 5.5 can therefore only be interpreted to a limited extent.

If the time that has passed since the start of the optimization process is also determined in each iteration, the norm of the gradient can be plotted against time. However, since both the norm of the gradient and the elapsed time differ in each iteration, it is no longer justified to calculate the arithmetic mean. Since the results differ clearly depending on the starting configuration, such averaging is necessary. Furthermore, the computing time is also influenced by background processes. Therefore, despite the disadvantages mentioned, the iteration number was used instead of the elapsed time.

5.4   Influence of Various Parameters on the Particle Distance

Since Sect. 5.3 indicates that (RGD) leads to the most accurate results among the methods considered, it is used to examine the relationship between the average particle distance Δr and other quantities like the momentum p, the plasma wavelength

λ and the number of electrons N. The parameter selection determined in Sect. 5.2 is used and the number of iterations is increased to 2000 for additional accuracy. The numerically obtained scaling laws can then be compared with the results from [17].

As derived analytically in [17], the average particle distance Δr of the electrons in the equilibrium state scales like

$$\Delta r \propto p^{-1/3}\lambda^{2/3}, \tag{5.10}$$

where p is again the longitudinal particle momentum and λ the plasma wavelength. This confirms the plausibility of our results presented in Fig. 5.6.

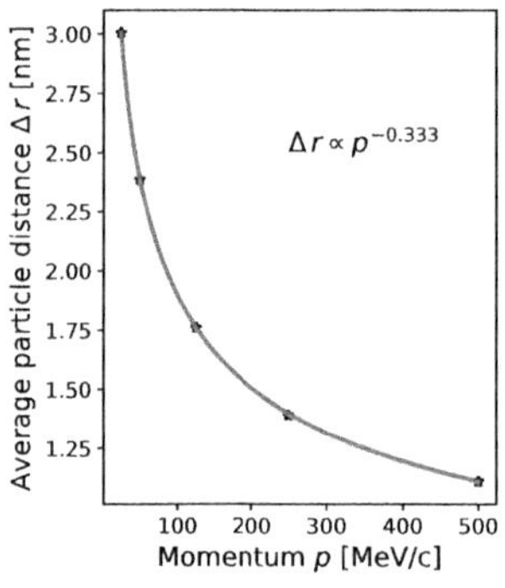

(a) Average particle distance depending on the momentum while $\lambda = 0.01\,\mathrm{cm}$ and $N = 100$.

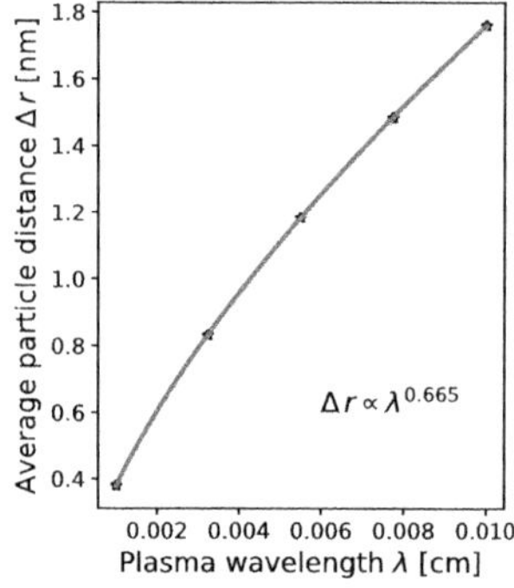

(b) Average particle distance depending on the wavelength while $p = 125\,\mathrm{MeV/c}$ and $N = 100$.

Figure 5.6 Average particle distance for different values of the momentum and the wavelength and an exponential fit in each case

For the number of particles N, the relationship with the average particle distance is somewhat more complicated. Unfortunately, the choice (5.9) turned out to be too rough when very small numbers of particles are considered. For such particle numbers, the optimal parameter selection seems to depend particularly strongly on the start configuration, so that no generally valid recommendation is possible. For this reason, additional effort was made to select the best possible parameters for $N \in \{2, 3, \ldots, 11\}$ for specific start configurations. The procedure was again to use a grid search to find start parameters for `min_f` and then to use `min_f` to identify those parameters for which the norm of the gradient is as small as possible.

As it can be seen from Table 5.1 and Fig. 5.8a, for small numbers of particles, no clear correlation is recognizable.

Table 5.1 Average particle distance Δr for small values of the particle number N while $p = 125\,\mathrm{MeV/c}$ and $\lambda = 0.01\,\mathrm{cm}$

N	2	3	4	5	6	7	8	9	10	11
$\Delta r/\mathrm{nm}$	2.25	2.58	2.49	2.34	2.19	2.53	2.35	2.21	2.16	2.26

Figure 5.7 can be helpful finding an explanation for the observed values. For $N \in \{3, \ldots, 6\}$ particles we get a regular polygon with N edges. If one imagines a circle on which the particles are located approximately, the radius increases with increasing N, but as the circle becomes more and more densely occupied with particles, the average particle distance Δr decreases for increasing N, as it can be seen in Table 5.1. The fact that the lowest value was determined for the hexagonal arrangement is consistent with Kepler's conjecture, which states that the highest density can be achieved by this structure.

By adding a seventh electron, something remarkable happens: The seventh particle is placed in the center of the hexagon. The repulsive electromagnetic force between the center and the edge significantly increases the radius of the corresponding circle. This explains the sharp jump in the average particle distances from $N = 6$ to $N = 7$ in Table 5.1.

Considering $N \in \{7, 8, 9\}$ particles, one electron is located in the center and the remaining $N - 1$ electrons form a regular polygon with $N - 1$ edges, so that the average particle distance decreases again for increasing N.

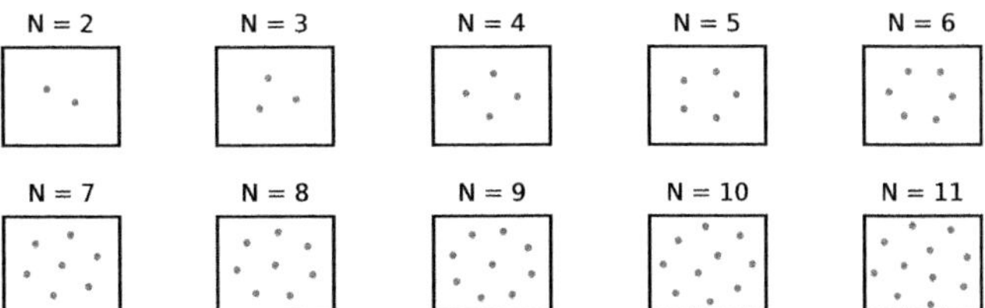

Figure 5.7 Final electron positions for different values of the particle number while $p = 125\,\mathrm{MeV/c}$ and $\lambda = 0.01\,\mathrm{cm}$

In the following, we use again the parameters recommended in (5.9). By increasing the number of particles, a tendency is slowly becoming apparent (see Fig. 5.8b).

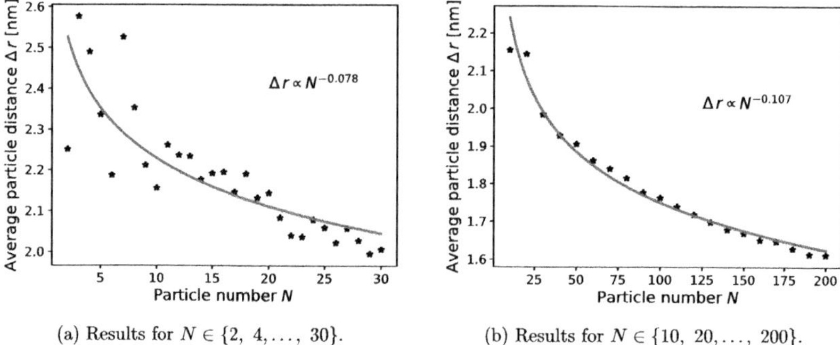

(a) Results for $N \in \{2,\ 4, \ldots,\ 30\}$.   (b) Results for $N \in \{10,\ 20, \ldots,\ 200\}$.

Figure 5.8 Average particle distance for small values of N while $p = 125\,\mathrm{MeV/c}$ and $\lambda = 0.01\,\mathrm{cm}$

Nevertheless, for sufficiently large numbers of particles an exponential fit is justified and our results (see Fig. 5.9) agree with the in [17] numerically observed scaling law $\Delta r \propto N^{-0.15}$.

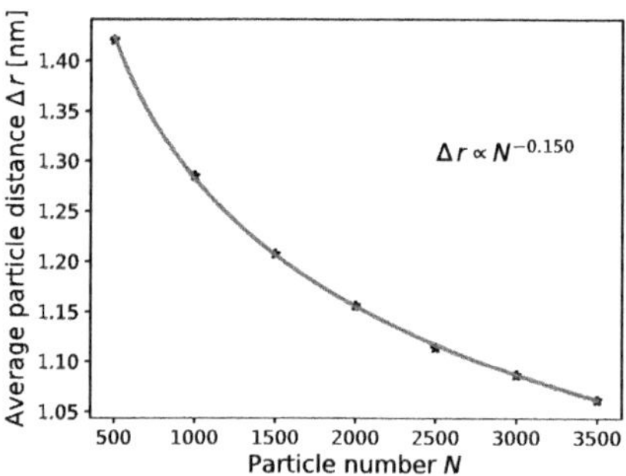

Figure 5.9 Average particle distance for different values of the particle number while $p = 125\,\mathrm{MeV/c}$ and $\lambda = 0.01\,\mathrm{cm}$ and an exponential fit

Conclusion and Outlook 6

This thesis started with some crucial properties of plasma like quasi-neutrality and the dispersion relation. The plasma frequency, the electron temperature, the electron density and other quantities help to distinguish between different types of plasma. Furthermore, the ponderomotive force and the equations of motion for the relativistic Lagrangian were derived. With these discussed principles we were able to describe how a laser pulse can create plasma cavities with strong electric fields, which can be used to accelerate charged particles like electrons. In order to describe the total energy of the electrons in the witness beam, we derived the Hamiltonian in accordance with [16]. This derivation is based on a phenomenological model [10], that uses rather simplifying assumptions. Beyond that, the electron-electron interactions were modeled using the full Liénard-Wiechert potentials. To examine the equilibrium state of the electrons in the witness beam, those particle positions that minimize the Hamiltonian are sought. For the numerical minimization of the Hamiltonian, [16] used the Gradient Descent Method (GDM) with Armijo Line Search. The aim of this thesis was to examine to what extent (GDM) is suitable for this purpose, and whether modifications of (GDM) can reduce the computational effort and increase the accuracy.

After a detailed discussion of the disadvantages of the previously used (GDM), three modifications of it were presented, namely the Heavy Ball Method (HBM), Nesterov's Accelerated Gradient Descent (NAG) and the Relativistic Gradient Descent (RGD). We applied our algorithms to the two-dimensional problem to minimize the Hamiltonian describing electrons in the witness beam on a slice perpendicular to the direction of propagation of the laser pulse. It was determined how the parameters occurring in the methods under consideration should be selected in our setting in order to obtain the best possible results. The procedure was to first determine rough parameters using a grid search and then to use these roughly

M. Hagedorn, *Minimization Problems for the Witness Beam in Relativistic Plasma Cavities*, BestMasters, https://doi.org/10.1007/978-3-658-46226-0_6

determined parameters as starting values for a finer optimization using the function `min_f`. The performance of the four algorithms was compared and the analysis indicated that (RGD) seems to be the most suitable method in our case, but it should be mentioned that the variance is quite large. Numerical scaling laws were derived using (RGD) to determine how the average particle distance is related to the momentum, the plasma wavelength and the particle number respectively. The scaling laws obtained were in good agreement with those already known from [17].

A future project is to examine how (RGD) performs in the three-dimensional case. This is particularly interesting because relativistic effects will then have a greater influence. This is because an electron moving at almost the speed of light cannot perceive an electron following the first one in its light cone.

One mathematical model using a Taylor expansion to approximate the Liénard-Wiechert potentials was suggested in [11]. In [62] the calculation of implicit retarded times was avoided by using a Lorentz transformation of the electromagnetic fields. Depending on the number of particles, the particles line up one after the other, begin to twirl up into a helix-like structure and finally form a shell-like structure. Interestingly, [62] describes that some patterning becomes recognizable on the formation on shells for a high precision of (GDM). With (RGD) we can hopefully examine this patterning in more detail in a subsequent work.

References

1. Stefaan Tavernier. *Experimental techniques in nuclear and particle physics.* Springer Nature, 2010.
2. Ragnar Hellborg and Göran Skog. Accelerator mass spectrometry. *Mass spectrometry reviews*, 27(5):398–427, 2008.
3. HIT. Heidelberger ionenstrahl-therapiezentrum, 2022. URL https://www.klinikum.uni-heidelberg.de/interdisziplinaere-zentren/heidelberger-ionenstrahl-therapiezentrum-hit.
4. CERN. European organization for nuclear research, 2022. URL https://public-archive.web.cern.ch/en/lhc/Facts-en.html.
5. Atlas Collaboration, Georges Aad, T Abajyan, B Abbott, J Abdallah, S Abdel Khalek, Ahmed Ali Abdelalim, O Abdinov, R Aben, B Abi, et al. A particle consistent with the higgs boson observed with the atlas detector at the large hadron collider. *Science*, 338(6114):1576–1582, 2012.
6. Peter W Higgs. Broken symmetries and the masses of gauge bosons. *Physical review letters*, 13(16):508, 1964.
7. Helmut Wiedemann and Helmut Wiedemann. Charged particle acceleration. *Particle Accelerator Physics II: Nonlinear and Higher-Order Beam Dynamics*, pages 163–228, 1995.
8. Victor Malka. Laser plasma accelerators. *Physics of Plasmas*, 19(5), 2012.
9. Alexancer Pukhov and Jürgen Meyer-ter Vehn. Laser wake field acceleration: the highly non-linear broken-wave regime. *Applied Physics B*, 74:355–361, 2002.
10. I Kostyukov, A Pukhov, and S Kiselev. Phenomenological theory of laser-plasma interaction in "bubble" regime. *Physics of Plasmas*, 11(11):5256–5264, 2004.
11. Johannes Thomas, Marc M Günther, and Alexander Pukhov. Beam load structures in a basic relativistic interaction model. *Physics of Plasmas*, 24(1), 2017.
12. FG Desforges, BS Paradkar, Magnus Hansson, J Ju, Lovisa Senje, TL Audet, A Persson, S Dobosz-Dufrénoy, Olle Lundh, G Maynard, et al. Dynamics of ionization-induced electron injection in the high density regime of laser wakefield acceleration. *Physics of Plasmas*, 21(12), 2014.
13. V Petrillo, A Bacci, R Ben Alì Zinati, I Chaikovska, C Curatolo, M Ferrario, C Maroli, C Ronsivalle, AR Rossi, L Serafini, et al. Photon flux and spectrum of γ-rays compton sources. *Nuclear Instruments and Methods in Physics Research Section A: Accelerators, Spectrometers, Detectors and Associated Equipment*, 693:109–116, 2012.

14. I Kostyukov, E Nerush, A Pukhov, and V Seredov. A multidimensional theory for electron trapping by a plasma wake generated in the bubble regime. *New Journal of Physics*, 12(4):045009, 2010.

15. Eugene Wigner. On the interaction of electrons in metals. *Physical Review*, 46(11):1002, 1934.

16. Lars Reichwein, Johannes Thomas, and Alexander Pukhov. Two-dimensional structures of electron bunches in relativistic plasma cavities. *Physical Review E*, 98(1):013201, 2018.

17. Lars Reichwein. *Struktur von coulomb-clustern im bubble-regime*. Springer, 2020.

18. Alexander Pukhov. *Lecture on Theoretical Plasma Physics*. HHU Düsseldorf, 2018/2019.

19. Karl-Heinz Spatschek. *Astrophysik*. Springer, 2003.

20. Karl-Heinz Spatschek. *Theoretische Plasmaphysik: Eine Einführung*. Springer-Verlag, 2013.

21. Ulrich Stroth. *Plasmaphysik*. Springer, 2011.

22. Giovanni Manfredi and Jérôme Hurst. Solid state plasmas. *Plasma Physics and Controlled Fusion*, 57(5):054004, 2015.

23. Paul Gibbon and Eckhart Förster. Short-pulse laser-plasma interactions. *Plasma physics and controlled fusion*, 38(6):769, 1996.

24. José A Bittencourt. *Fundamentals of plasma physics*. Springer Science & Business Media, 2004.

25. Chuanjun Huang and Laifeng Li. Magnetic confinement fusion: a brief review. *Frontiers in Energy*, 12:305–313, 2018.

26. R Betti and OA Hurricane. Inertial-confinement fusion with lasers. *Nature Physics*, 12(5):435–448, 2016.

27. ITER. International thermonuclear experimental reactor, 2022. URL https://www.iter.org/.

28. AB Zylstra, AL Kritcher, OA Hurricane, DA Callahan, JE Ralph, DT Casey, A Pak, OL Landen, B Bachmann, KL Baker, et al. Experimental achievement and signatures of ignition at the national ignition facility. *Physical Review E*, 106(2):025202, 2022.

29. Goetz Lehmann and Karl-Heinz Spatschek. Transient plasma photonic crystals for high-power lasers. *Physical review letters*, 116(22):225002, 2016.

30. G Lehmann and KH Spatschek. Laser-driven plasma photonic crystals for high-power lasers. *Physics of Plasmas*, 24(5), 2017.

31. G Lehmann and KH Spatschek. Plasma-based polarizer and waveplate at large laser intensity. *Physical Review E*, 97(6):063201, 2018.

32. Dieter Schardt, Thilo Elsässer, and Daniela Schulz-Ertner. Heavy-ion tumor therapy: Physical and radiobiological benefits. *Reviews of modern physics*, 82(1):383, 2010.

33. Stuart PD Mangles, CD Murphy, Zulfikar Najmudin, Alexander George Roy Thomas, JL Collier, Aboobaker E Dangor, EJ Divall, PS Foster, JG Gallacher, CJ Hooker, et al. Monoenergetic beams of relativistic electrons from intense laser–plasma interactions. *Nature*, 431(7008):535–538, 2004.

34. Matthias Bartelmann, Björn Feuerbacher, Timm Krüger, Dieter Lüst, Anton Rebhan, and Andreas Wipf. *Theoretische Physik*. Springer, 2015.

35. J. Eichler and H.J. Eichler. *Laser: Grundlagen · Systeme · Anwendungen*. Springer Berlin Heidelberg, 1991.

36. Donald A Gurnett and Amitava Bhattacharjee. *Introduction to plasma physics: With space, laboratory and astrophysical applications*. Cambridge University Press, 2017.

37. Philip McCord Morse and K Uno Ingard. *Theoretical acoustics*. Princeton university press, 1986.

38. Robert J Goldston. *Introduction to plasma physics*. CRC Press, 2020.

39. Francis F Chen et al. *Introduction to plasma physics and controlled fusion*, volume 1. Springer, 1984.

40. Levan N Tsintsadze, Kunioki Mima, and Kyoji Nishikawa. Overdense propagation of a relativistically intense laser light. *Plasma physics and controlled fusion*, 40(11):1933, 1998.

41. Roland Fischer, KM Hanson, V Dose, and W von Der Linden. Background estimation in experimental spectra. *Physical Review E*, 61(2):1152, 2000.

42. Natanael Karjanto. Bright soliton solution of the nonlinear schrödinger equation: Fourier spectrum and fundamental characteristics. *Mathematics*, 10(23):4559, 2022.

43. Andrea Macchi. *A superintense laser-plasma interaction theory primer*. Springer Science & Business Media, 2013.

44. Harry Moritz Schey. Div, grad, curl, and all that. *Div*, 1973.

45. Brice Quesnel and Patrick Mora. Theory and simulation of the interaction of ultraintense laser pulses with electrons in vacuum. *Physical Review E*, 58(3):3719, 1998.

46. Louis N Hand and Janet D Finch. *Analytical mechanics*. Cambridge University Press, 1998.

47. Toshiki Tajima and John M Dawson. Laser electron accelerator. *Physical review letters*, 43(4):267, 1979.

48. Pisin Chen, John M Dawson, Robert W Huff, and Thomas Katsouleas. Acceleration of electrons by the interaction of a bunched electron beam with a plasma. *Physical review letters*, 54(7):693, 1985.

49. Allen Caldwell, Konstantin Lotov, Alexander Pukhov, and Frank Simon. Proton-driven plasma-wakefield acceleration. *Nature Physics*, 5(5):363–367, 2009.

50. Toshiki Tajima, XQ Yan, and T Ebisuzaki. Wakefield acceleration. *Reviews of Modern Plasma Physics*, 4:1–72, 2020.

51. Henrik Ekerfelt, Martin Hansson, Isabel Gallardo González, Xavier Davoine, and Olle Lundh. A tunable electron beam source using trapping of electrons in a density down-ramp in laser wakefield acceleration. *Scientific Reports*, 7(1):12229, 2017.

52. P Muggli. Beam-driven, plasma-based particle accelerators. *arXiv preprint* arXiv:1705.10537, 2017.

53. Andreas Malcherek. *Elektromagnetismus und Gravitation: die Vereinheitlichung der klassischen Physik*. Springer, 2022.

54. Lev Davidovič Landau and Evgenii M Lifschitz. *Klassische Feldtheorie*. De Gruyter, 1963.

55. Jon Lee and Sven Leyffer. *Mixed integer nonlinear programming*, volume 154. Springer Science & Business Media, 2011.

56. Yurii Nesterov. *Introductory lectures on convex optimization: A basic course*, volume 87. Springer Science & Business Media, 2003.

57. Boris T Polyak. Some methods of speeding up the convergence of iteration methods. *Ussr computational mathematics and mathematical physics*, 4(5):1–17, 1964.

58. Guilherme França, Jeremias Sulam, Daniel Robinson, and René Vidal. Conformal symplectic and relativistic optimization. *Advances in Neural Information Processing Systems*, 33:16916–16926, 2020.

59. Markus Lazar and Florian Jarre. Calibration by optimization without using derivatives. *Optimization and Engineering*, 17:833–860, 2016.

60. GitHub repository. https://github.com/mhagedorn/min_f 2024.

61. Florian Jarre and Josef Stoer. Lineare programme, beispiele und definitionen. *Optimierung: Einführung in mathematische Theorie und Methoden*, 2019.

62. Lars Reichwein, Johannes Thomas, and Alexander Pukhov. The filamented electron bunch of the bubble regime. *Laser and Particle Beams*, 38(2):121–127, 2020.